# Advances in Geographical and Environmental Sciences

Advances in Geographical and Environmental Sciences synthesizes series diagnostigation and prognostication of earth environment, incorporating challenging interactive areas within ecological envelope of geosphere, biosphere, hydrosphere, atmosphere and cryosphere. It deals with land use land cover change (LUCC), urbanization, energy flux, land-ocean fluxes, climate, food security, ecohydrology, biodiversity, natural hazards and disasters, human health and their mutual interaction and feedback mechanism in order to contribute towards sustainable future. The geosciences methods range from traditional field techniques and conventional data collection, use of remote sensing and geographical information system, computer aided technique to advance geostatistical and dynamic modeling.

The series integrate past, present and future of geospheric attributes incorporating biophysical and human dimensions in spatio-temporal perspectives. The geosciences, encompassing land-ocean-atmosphere interaction is considered as a vital component in the context of environmental issues, especially in observation and prediction of air and water pollution, global warming and urban heat islands. It is important to communicate the advances in geosciences to increase resilience of society through capacity building for mitigating the impact of natural hazards and disasters. Sustainability of human society depends strongly on the earth environment, and thus the development of geosciences is critical for a better understanding of our living environment, and its sustainable development.

Geoscience also has the responsibility to not confine itself to addressing current problems but it is also developing a framework to address future issues. In order to build a 'Future Earth Model' for understanding and predicting the functioning of the whole climatic system, collaboration of experts in the traditional earth disciplines as well as in ecology, information technology, instrumentation and complex system is essential, through initiatives from human geoscientists. Thus human geoscience is emerging as key policy science for contributing towards sustainability/survivality science together with future earth initiative.

Advances in Geographical and Environmental Sciences series publishes books that contain novel approaches in tackling issues of human geoscience in its broadest sense-books in the series should focus on true progress in a particular area or region. The series includes monographs and edited volumes without any limitations in the page numbers.

Fayez Alaily

# Ecosystems of Extremes

## Climate, Soils, and Sustainable Land Use in the Central-Eastern Sahara

Springer

Fayez Alaily
Institute of Ecology
Technical University of Berlin
Berlin, Germany

ISSN 2198-3542 ISSN 2198-3550 (electronic)
Advances in Geographical and Environmental Sciences
ISBN 978-3-031-76909-2 ISBN 978-3-031-76910-8 (eBook)
https://doi.org/10.1007/978-3-031-76910-8



This Springer imprint is published by the registered company Springer Nature Switzerland AG
The registered company address is: Gewerbestrasse 11, 6330 Cham, Switzerland

# Preface and Acknowledgements

From 1981 to 1986, an interdisciplinary research project was carried out to investigate geoscientific problems of the Central East Sahara in southwest Egypt. The research project was titled "Geoscientific Problems in Arid Areas". In addition to soil scientists, 14 scientific disciplines were involved in this project: mineralogy, photogrammetry, paleontology, sedimentology, remote sensing, geophysics, petrology, geochemistry, botany, hydrology, geology, physical geography, and mining sciences. Without contacts with such disciplines and their findings, the present monograph would not have been possible. The project was equipped with a truck for water and fuel, an Unimog with excavator, and 5 off-road cars. On our trips we had no GPS and no reliable topographic maps. We could only navigate with the help of satellite images in scale 1:500,000 and a compass. Five expeditions, each 4–8 weeks, were carried out from 1981–1986 during the winter months. For the safety of the group, we had to send our positions to the General Petroleum Company every evening. This was not always possible for various reasons.

The aim of our work was to produce soil maps for an area which was not processed by soil investigators. In addition, the FAO-UNESCO soil classification system of 1974 should be tested in an extremely dry region. Based on the results of our investigations, the "World Reference Base for Soil Resources (WRB)" was elaborated, and, in particular, the desert soils were better classified.

The soil maps were printed at the Technical University of Applied Sciences Berlin under the direction of Prof. Dr. G. Pöhlmann with precise coordinates. The laboratory analyses were carried out at the Institute of Ecology of the Technical University of Berlin. The chemotechnicians Mrs. Karin Mayer, Mrs. Diane Zapf, Mrs. Christina Ehrlicher, and Mrs. Sabine Dumke were responsible for carrying out the laboratory analyses. I would like to thank them for analytical work of a large number of soil and rock samples.

Special thanks go to Prof. Dr. E. Klitzsch: the initiator of the interdisciplinary research project "Geoscientific Problems in Arid Areas" and Prof. Dr. H. P. Blume: for intensive and fruitful discussions about soil genesis and land use. I also would like to thank Prof. Dr. R. Bornkamm and Priv. Doz. Dr. H. Kehl for interesting ecological and plant physiological discussions, Dr. J. Pohlmann for the digital interpretation of satellite images, and Prof. Dr. H. Meshref (Mansoura University) for his intensive help in the field.

For financial support of this research, the Deutsche Forschungsgemeinschaft (DFG) is gratefully acknowledged. Our thanks are extended to the Egyptian "General Petroleum Company" (GPC) for logistical support of customs formalities in Egypt and for the import of vehicles and spare parts as well as for the export of soil and rock samples.

In this book I have presented most of my experiences and professional results in the desert. It should assist students and scientists working in the Sahara or other desert regions with their studies or their work. Almost all field observations and laboratory results are illustrated using diagrams, sketches, and photos. The heart of the Sahara was first published in 2008 in German (das Herz der Sahara) by Projekte Verlag. The current second edition is written in English and contains some improvements and supplements. The goal of this edition is mainly to reach a larger number of readers, especially those in countries of the arid zone.

Berlin, Germany Fayez Alaily

# Contents

# Chapter 1
# Introduction

**Abstract** The investigated region is found in Central East Sahara (SW Egypt) under the tropic of Cancer and belongs to so-called horse-latitude deserts. In addition, the great distance from the Atlantic and the Mediterranean Sea makes the area the driest region on earth. This region has been little researched scientifically and has never been investigated by soil scientists. The results of our soil investigation served to refine the soil classification of the World Soil Map (FAO-UNESCO in Soil map of the world (1:5,000,000), vol I, Legend, UNESCO, Paris, 1974) for full and extreme deserts. In the FAO soil classification system of 1988 and the "World Reference Base" (ISSS, ISRIC, FAO in World reference base for soil resources. FAO, World Soul Resources Report, No. 84, Rome, 1998), basic results from our investigations have been used, in particular for the definition of the "yermic phase" or the "yermic-" and "aridic properties".

**Keywords** World reference base · Horse-latitude deserts

The arid soil zone is one of the largest soil zones on earth. According to Thorne and Peterson (1954), the arid soil zone covers 25% of the earth's surface. There are 28 countries within this soil zone. Apart from the oil-producing countries, most of the countries in this zone are among the poorest countries in the world. Their poverty is primarily due to the dry climate. Through soil studies and knowledge of the function of the ecosystems in these areas, a concept for optimizing land use can be developed.

From all the desert regions of the world, the Sahara has the largest contiguous area and occupies about 30% of the surface of the African continent. It is the largest tropical dry desert on earth, with an area of 8.6 million $km^2$. The term "Sahara" means desert in Arabic. The Arabic name for the Sahara is "Al Sahara El Kubra"; literally translated it means "the largest desert". It stretches from the Atlantic coast to the Red Sea (4500–5500 km west–east) and from the Mediterranean coast or from the Atlas Mountains to the Sahel (1500–2000 km north–south). To the north are Egypt, Libya, and the Magreb countries (Tunisia, Algeria, and Morocco). To the south are Sudan, Chad, Niger, Mali, Burkina Faso, and Mauritania. In Egypt, the Sahara is interrupted by the Nile. To the west of the Nile valley, the Sahara in Egypt is called the Western

 
F. Alaily, *Ecosystems of Extremes*, Advances in Geographical and Environmental Sciences, https://doi.org/10.1007/978-3-031-76910-8_1

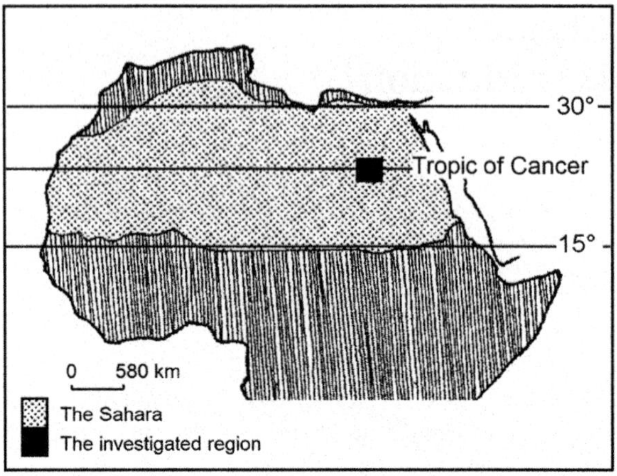


**Fig. 1.1** Situation of the Sahara and the investigated region

Desert (Nubian or Libyan Desert) and eastern the Nile valley it is called the Eastern Desert.

The Sahara forms the western part of the Afro-Asian desert belt. It is between the 15 and 30 latitudes and is divided into two halves by the tropic of Cancer (Fig. 1.1). Thus, the Sahara belongs to the turning circle deserts or the so-called horse-latitude deserts. In general, precipitation decreases from the northern and southern borders of the Sahara toward the tropic of Cancer and from west to east, so that the driest part of the Sahara can be found in the east at the turning circle in southwest Egypt, in the region of our investigation. This area is located between the 23rd and 26th latitudes and between the 26th and 31st longitude. It is located in southwest Egypt and comprises most of the New Valley Governorate. It borders Sudan to the south and Libya to the west. The most famous locations in the area are the Oases Dakhla and Kharga, as well as Tarfawi, Gilf Kebir, and the Great Sand Sea (Fig. 1.2).

Until we started our investigations 1981, the region has been mainly studied by geographers and geologists (Busche 1998; Rautenstrauch-Joest-Museum 1978) but not by soil scientists. As soil maps of Eastern Sahara, only the soil map of Egypt on a scale of 1:4,000,000 (El Gabaly et al. 1969a and 1969b) and the FAO world soil map on a scale of 1:5,000,000 (FAO-UNESCO 1977) existed. However, the maps are mostly not based on ground surveys. The latter are only available for parts of the Kharga–Dakhla Oases Complex (Beller Consult 1985) and as selective surveys in East Uweinat (Hassan 1979). They were done using simple methods to determine land use worthiness. The fertility of some soils of the Central Sahara (in Algeria, Libya, and Egypt) was described by Vogg (1981). The soil conditions of the Fezzan in southwest Libya are also well known (Petermann and Blume 1979; Petermann 1988). Our detailed soil surveys were carried out in the region of Central Eastern Sahara in Western Egypt as part of an interdisciplinary project funded by the German Research Foundation. The results of the soil investigation served to refine the soil classification of the World Soil Map (FAO-UNESCO 1974) for full and extreme deserts. In the FAO soil classification system of 1988 and the "World Reference

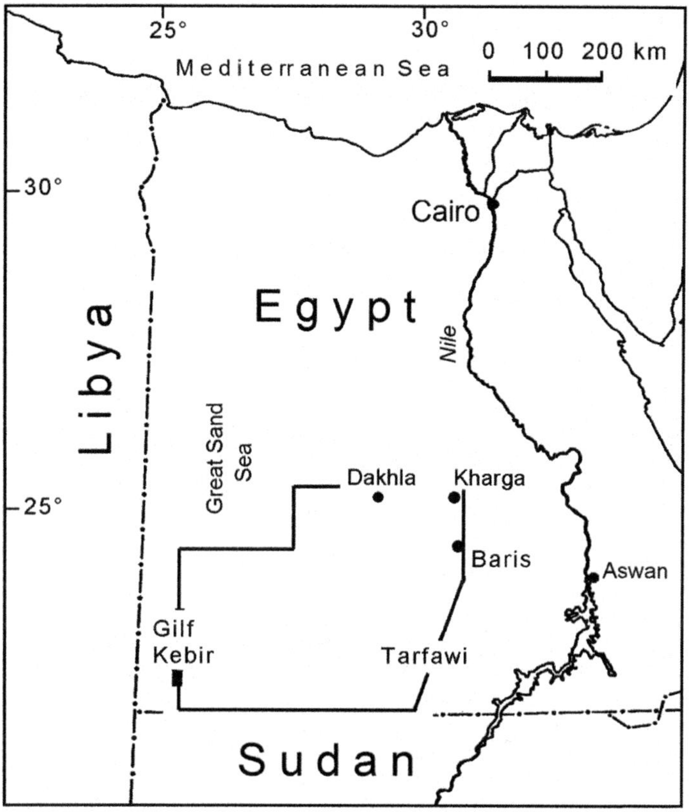


**Fig. 1.2** Location of the investigated region

Base" (ISSS, ISRIC, FAO 1998), basic results from our investigations have been used, in particular for the definition of the "yermic phase" or the "yermic-" and "aridic properties".

From the field and laboratory work, a soil association map (Alaily and Blume 1986), a vegetation map (Bornkamm and Kehl 1987), and a map of land suitability for irrigated agriculture (Alaily 1990) in scale 1:1,000,000 as well as numerous publications were done. The publications focus on soil genesis and soil assessment for irrigated arable land. Also detailed soil investigations as well as large-scale soil maps of small areas at Bir Tarfawi, Bir Safsaf and the Bulaq oasis were made and published by Smettan (1987), Ruck (1989) and Blume et al. (1987).

The present book summarizes all results obtained and their significance for soil sociology, genesis, ecology, and potential uses. In addition, all observed phenomena that have not yet been scientifically clarified, as well as experiences and adventures from these expeditions, are presented. Locations that are frequently mentioned in this book are shown in the appendix.

## References

Alaily F (1990) Ecological maps of the Western Desert of Egypt; land suitability map 1:1000000. TFH-Berlin

Alaily F, Blume H-P (1986) Ecological maps of the Western Desert of Egypt; soil association map 1:1000000. TFH-Berlin

Beller Consult Ltd (1985) Water resources and soil potential development project New Valley. Freiburg/Siegen

Blume H-P, Pfisterer U, Lange H (1987) Bodenverhältnisse des BirSafsaf in SW Ägypten. Berliner geowissenschaftl. Abh. A 78:111–116

Bornkamm R, Kehl H (1987) Ecological maps of the Western Desert of Egypt; Vegetation unit map 1:1000000. TFH-Berlin

Busche D (1998) Die Zentral Sahara. Justus Perthes Verlag, Gotha, p 284

El Gabaly MM, Gewaifel IM, Hassan MN, Rozanov BG (1969a) Soils and soil regions of U. A. R. Alexandria Univ Res Bull 21

El Gabaly MM, Gewaifel IM, Hassan MN, Rozanov BG (1969b) Soil map and land resources of U. A. R. Alexandria Univ Res Bull **22**

FAO-UNESCO (1974) Soil map of the world (1:5000000), vol I, Legend, UNESCO, Paris

FAO-UNESCO (1977) Soil map of the world, vol VI, Africa, 300 p

FAO-UNESCO (1988) Soil map of the world, 1:5000000, Revised Legend. Soil Res Rep 60:119

Hassan N (1979) Preliminary investigations of groundwater and soil resources in East Oweinat Area, Western desert Egypt, Cairo

ISSS, ISRIC, FAO (1998) World reference base for soil resources. FAO, World Soul Resources Report, No. 84, Rome

Petermann Th (1988): Böden des Fezzan/Südwestlibyen – Studie zur Klassifikation, Vergesellschaftung, Ökologie und Nutzungseignung von Böden einer Großregion der Sahara. Schriftenreihe Inst. Pflanzenernähr. Bodenkunde Univ. Kiel, Nr. 1

Petermann Th, Blume H-P (1979) Yermosol-Solonchak-Bodenschaft des Fezzan, II. Soziologie und Bewässerungseignung.- Mitteilgn. Dtsch. Bodenkundl. Gesellsch. 29:811–820

Rautenstrauch-Joest-Museum (1978) Sahara: 10000 Jahre zwischen Weide und Wüste. Museen der Stadt Köln, p 470.

Ruck A (1989) Experimentelle Ökodynamik an zwei Standorten der Extremwüste. Diss., Fachbereich 14, TU-Berlin, p 280

Smettan U (1987) Typische Böden und Bodengesellschaften der Extremwüste Südwest-Ägyptens. Berliner geow. Abh. A 83:1–190

Thorne D, Peterson H (1954) Irrigated soils—their fertility and management. The Blakiston Comp. Inc., New York

Vogg R (1981) Bodenressourcen arider Gebiete. Stuttgarter Geographische Studien 97:224

# Chapter 2
# Climate

**Abstract** Precipitation is 0.5 mm/a. Rainfalls occur mostly in winter and sometimes in summer in a few days after several decades of drought. The average relative humidity of the air is 35%, with a maximum of only 48% in December and a minimum of 28% in June and July. Dew events are very seldom. In the Gilf Kebir, there may have been a 20- to 30-year rainy period in the last 200 years. The average annual air temperature is 23 °C, with an average minimum of 11 °C in January and an average maximum of 30 °C in June, July, and August. Frost can occur in January and February. The upper 40 cm of the soil are affected by daily temperature fluctuation (< 10 to > 40 °C), and the lower parts are affected by only annual temperature changing. About 6.3% of the year, the area is hit by sandstorms that reshape dunes, set them in motion, and rub stones by sand blower that occur on the soil surface.

**Keywords** Precipitation · Relative humidity · Dew · Air temperature · Frost · Sandstorm · Sand blower

The aridity of the region is mainly due to its location on the Tropic of Cancer and the great distance from the coast. Climatic data are very sparse and not comprehensive. The region has only three meteorological stations: in Dakhla, Kharga, and Tarfawi. The meteorological station of Tarfawi has been in operation since 1980 and is officially referred to as the "East Uweinat". The meteorological stations Dakhla and Kharga are located in the north and northeast border of the investigated region, and they are found within the oasis area. Therefore, they are not representative of the entire region. Moreover, there are no meteorological data for the plateau of the Gilf Kebir—which is the highest elevated location—as well as the Great Sand Sea with its large and dens amount of dunes.

 
F. Alaily, *Ecosystems of Extremes*, Advances in Geographical and Environmental Sciences, https://doi.org/10.1007/978-3-031-76910-8_2

## 2.1 Precipitation and Condensation

The location of the investigated region at the Tropic of Cancer means that it lies almost permanently in a climate zone influenced by meteorological high pressure, which is determined by the planetary air circulation (Fig. 2.1). It is characterized mainly by descending dry air mass, by a low cloud cover, and thus high radiation values. The area is rarely affected by precipitation and by the low air pressure from the inner-tropical convergence zone (ITC) or from the low air pressure of the middle latitudes connected with the circular west winds. Winter rains can occasionally occur and sometimes even into the southern part of the Sahara, but the summer rains of tropical origin (monsoons) penetrate only in exceptional cases over 16°–17° north latitude (Flohn 1982) and can therefore hardly influence the investigated area. The aridity caused by atmospheric high pressure is also favored by the great distance from the coast.

According to the Dakhla and Kharga Meteorological Stations, the average annual rainfall is only 0.8 mm and 0.5 mm, respectively (Fig. 2.2), that is the least measured annual precipitation on earth. Therefore, the studied region is considered to be the driest part of the world. The rarity of rainfall in southern Egypt explains why there was no word for rain in the Pharaonic language (Jung 2007). Rain was described in the new kingdom (1550–1000 BC) with the phrase "water of the sky". Rain was an undesirable natural event for the pharaohs, as it fell either in insignificant quantities or in heavy downpours, which caused a lot of damage. Photo 2.1 shows raindrop traces on a shard of clay and gives an impression of the intensity of episodic rainfall.

Figure 2.2 shows that the maximum daily rainfall measured over 45 years in the oases Dakhla and Kharga is 10–16 times higher than the average of a year. They are 8 mm in Dakhla and 7.5 mm in Kharga. This means that the rainfall will occur in a few days after several decades of drought. These rare rainfalls occur mostly in winter and occasionally in summer. Apart from the temporal variability, a spatial dispersion of precipitation must also be expected.

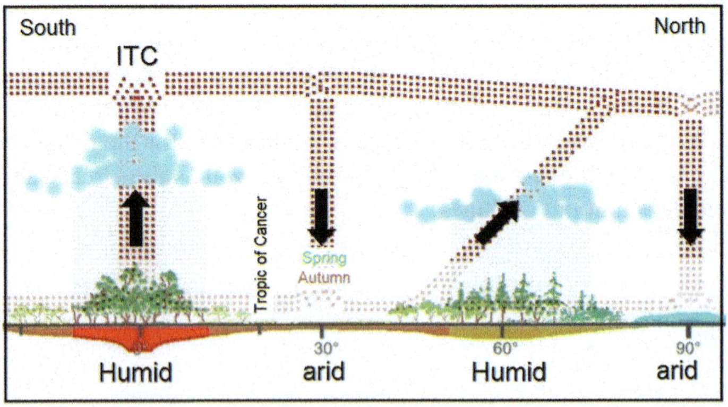


**Fig. 2.1** Global air circulation and global rain distribution

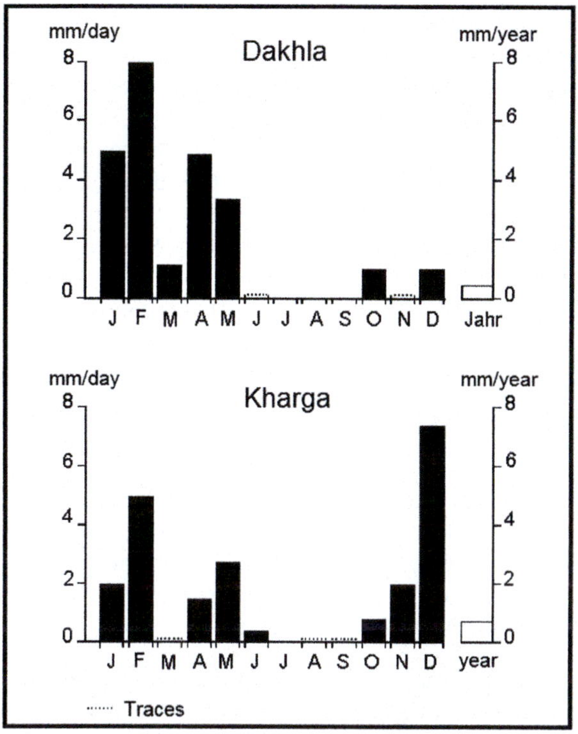


**Fig. 2.2** Maximum daily rainfall and average annual rainfall for the period 1931–1975

**Photo 2.1** Traces of raindrops that fell during an episodic rain event

Episodic rain penetrates the soil to a maximum depth of 10 cm and is responsible for all chemical reactions in the soil. They are meaningless for the existence of plants. During our expeditions (1982–1986), a single rainfall event of a few minutes took place on 25.01.1982 northeast of Wadi Contrast (North Gilf Kebir). The amount of water was sufficient to moisten about 1 cm depth of the soil sand cover.

In January and February 1982 and 1983, 6–7 days of thawing events were observed. The dew water was formed just before sunrise and only on the loose soil sand cover, which was moistened 1–2 mm deep. This water evaporated in a few minutes after sunrise. On stones of the soil cover, dew water was not observed. This must be explained by the high heat capacity of stones compared to the loose sand.

## 2.2 Last Episodic Wet Phase

Within the past centuries, a relative wet period with frequent episodic rain events has left traces in the sediments of Wadi Maftuh in Gilf Kebir. A buried and pseudo-carbonized tamarisk tree was found behind a barrier dune in the upper reaches of Wadi Maftuh (Photo 2.2 and Fig. 2.3). It was covered with five sandy layers, separated by four thin silty layers. Figure 2.3 shows that:

(1) the thickness of the sandy layers above the base of the tree trunk (layer 5) increases in direction of soil surface, whereas the thickness of the silty layers decreases in the same direction,
(2) the lowest silty layers in the root zone as well as the top two silty layers show takyric structure,

**Photo 2.2** Tamarisk plant buried by several sandy and silty layers

**Photo 2.3** Buried tamarisk tree in the upper reaches of Wadi Maftuh (eastern part of Gilf Kebir)

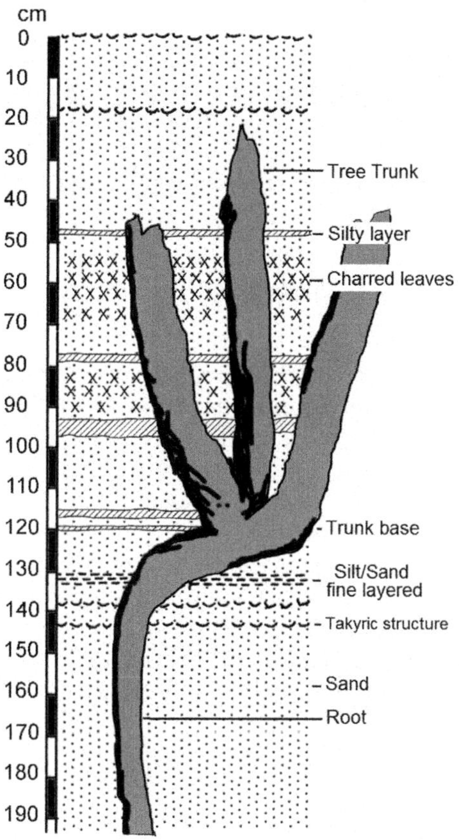


(3) the second and third layers (from the stem base upwards) are rich in carbonized tamarisk leaves.

The material of the silty layers can only be created by surface water. Therefore, the thickness of these layers is likely to reflect the amount of surface water or precipitation water at that time, while the thickness of the sandy layers can reflect the length of the dry period. This would mean that, during the last life phase of the tree, rainfall slowly decreased, and the length of the rain-free period increased. From the second sandy layer, water was scarce, and during the deposit of the third layer, water in the soil was probably no longer sufficient to keep the tree alive. Therefore, the tree died and dropped its last leaves during the accumulation of the third layer (counted from the tree trunk base upwards).

$^{14}$C-analysis of the tree trunk and roots gave an age of less than 200 years. Similar results were found in the $^{14}$C dating of plant roots derived from Bg horizons of a soil profile in the central area south of Tarfawi. This means that over the last two centuries at least one humid period has occurred, which provided enough water for the growth of the tamarisk tree. During this time, enough phytomass may have been produced to feed wild goats in Gilf Kebir.

The age of the tamarisk tree can be estimated from the thickness of its trunk, which roughly equates the duration of the wet period. The thickness of the trunk is 15–18 cm in diameter and could have taken 20–30 years to develop. This could also mean that this wet period lasted for around 20 years.

## 2.3 Air Humidity

The trade winds that blow through this arid zone are evaporative winds. The average relative humidity of the air is 35% (Fig. 2.4), with a maximum of only 48% in December and a minimum of 28% in June and July. At Bir Tarfawi, on the border with Sudan, relative humidity of less than 7% is measured in August, in several consecutive days. Even relative humidity of 4% is not uncommon. The extremely low humidity and the rarity of rainfall make the area the driest of the world.

According to the relative humidity, the potential evaporation is extremely high. It is 2068 mm for June, July, and August and 5837 mm in the year. The high dryness of the climate causes carcasses of animals to dry out in a very short time and are thus mummified naturally (Photos 2.3 and 2.4). Moreover, the oxidation and weathering of objects, e.g., made of iron, is strongly inhibited (Photo 2.5).

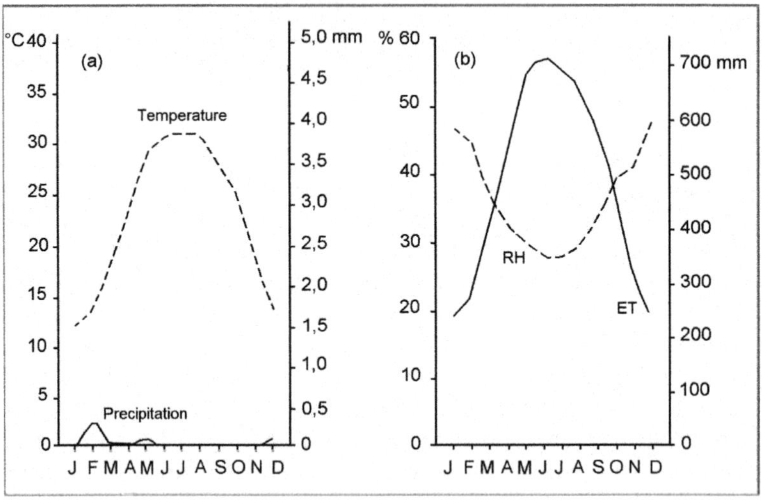


**Fig. 2.4** **a** Temperature and precipitation as well as **b** relative humidity (RH) and potential evaporation (ET) according to the Dakhla Meteorological Station (period 1931–1975)

**Photo 2.3** High dryness of the climate causes carcasses of animals to dry out within a short time and be naturally mummified

**Photo 2.4** Dry cow head (from Photo 2.3) on the asphalt road about 70 km west of the oasis Mut near Dakhla

## 2.4 Air Temperature

According to Dakhla Meteorological Station, the average annual air temperature is 23 °C, with an average minimum of 11 °C in January and an average maximum of 30 °C in June, July, and August (Fig. 2.4).

**Photo 2.5** For more than 60 years, a truck left over from the last world war (above), still without rust and with undamaged ammunition load (below). The truck belonged to a Canadian force and was found about 40 km west of Black Hill

Because of clear sky, the air temperature decreases noticeably at night and frost can occur in winter (Photo 2.6). At the meteorological station of East Uweinat (50 km south of Bir Tarfawi) from 1982 to 1985 frost days were measured on only one day in February 1982 and on 16 days in January and February 1983. In 1984 and 1985, no frost has been recorded. Also, in Kharga Oases frost can occur in January and February, at the Dakhla Oases from December until March. On January 17, 1982 at 6:00 O´clock am our team measured – 7 °C halfway from Tarfawi to Gilf Kebir.

**Photo 2.6** Sleeping bag with Hoarfrost of an early morning in November 1982 in Wadi Mashi at the Gilf Kebir

In Table 2.1, the monthly number of days with temperature minima < 5 °C and the number of days with temperature maxima of > 30 °C and > 40 °C are presented. These figures show that the highest air temperatures occur in summer, especially in June and July, and the lowest occur in winter especially in December and January. The average daily air temperature fluctuation is 17 °C. This temperature fluctuation applies to the air layer at a height of 2 m, where the meteorological data are collected. On the ground surface, the variation is much greater. The annual and the daily temperature fluctuations are the cause of many physical processes in soil.

Due to the big difference of temperature between the air layer close to the ground and the upper air layers, a mirage of objects that are far away can be seen (Photo 2.7). This phenomenon is often observed at midday on level ground, and it is explained by the relationships between air temperature, density, and refraction of light. It appears where two air layers with very different temperatures and densities are formed, and there the light rays are refracted or totally reflected at the density limit.

**Table 2.1** Number of days with air temperatures < 5 °C, > 30 °C, and > 40 °C (after the meteorological station of "East Uweinat", for the years 1982–1985)

| | J | F | M | A | M | J | J | A | S | O | N | D |
|---|---|---|---|---|---|---|---|---|---|---|---|---|
| *Days with temperature* | | | | | | | | | | | | |
| < 5 °C | 19.5 | 9.8 | 4.5 | 0 | 0 | 0 | 0 | 0 | 0 | 0 | 1.5 | 15.5 |
| > 30 °C | 2.0 | 1.3 | 9.3 | 22.4 | 30.0 | 30.0 | 31.0 | 31.0 | 30.0 | 23.3 | 2.8 | 0.5 |
| > 40 °C | 0 | 0 | 0 | 2.6 | 5.0 | 9.3 | 7.5 | 3.7 | 5.0 | 1.0 | 0 | 0 |

**Photo 2.7** Mirage or aerial reflection of hills. Such observations are made usually at midday

All data in Fig. 2.4 and Table 2.1 are influenced by the oasis microclimate. It is expected that the air temperature minima and the number of days with low temperatures outside the oases are even more frequent. Due to the altitude of the Gilf Kebir (500–1000 m. a. s. l.), the lowest temperatures are likely to occur in this part of the investigated region.

## 2.5 Soil Temperature

Figure 2.5 shows the average soil temperatures determined by the Dakhla meteorological station in 5, 10, 20, 50, and 100 cm depth, at 0, 6, 12, and 18 o'clock for the coldest and warmest months. Similar results have been obtained from the meteorological station of "East Uweinat".

The daily soil temperature fluctuations in 5, 10, 20, and 50 cm depth of a soil at the meteorological station "East Uweinat" are on average 11.4, 6.2, 2.1, and 0.5 °C, and the average annual temperature fluctuations are 39.2, 28.8, 25.0, and 18.6 °C (Fig. 2.6). These averages were taken from measurements at 8, 14, and 20 o'clock at the East Uweinat station. Since the lowest surface temperature is expected just before sunrise (between 4 and 6 o'clock), the absolute temperature fluctuations in the top centimeters of the soil must be greater than those shown in Fig. 2.6. During daytime, temperature fluctuations of 30 and 50°C in the top centimeters of the soil were measured by Meckelein (1959) in the Fezzan and Briem (1977) in the Tibesti.

According to the daily and annual soil temperature fluctuations at the meteorological stations "East Uweinat" and Dakhla, the soil can be divided into three layers (Fig. 2.6):

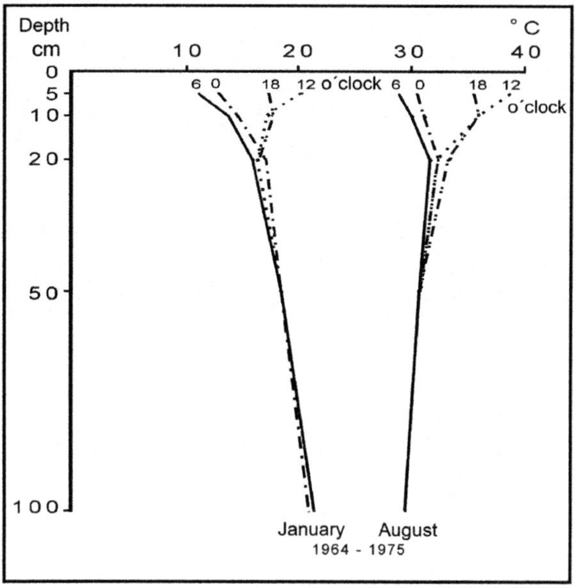


**Fig. 2.5** Average daily temperature fluctuations in the soil during the coldest (January) and warmest (August) months (1964–1975) at the Dakhla Meteorological Station

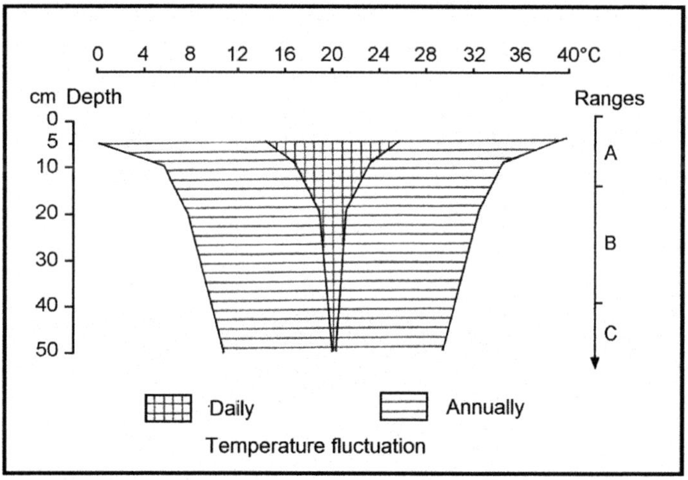


**Fig. 2.6** Depth function of soil temperature fluctuation-amplitude at Tarfawi for 1982 (data from the meteorological station "East Uweinat)

A. Upper layer, 0–15 cm depth; with high daily and annual temperature fluctuations.
B. Medium layer, 15–40 cm depth; with weak daily and high annual temperature fluctuations.
C. Lower layer, 40–> 100 cm depth; with no daily and with weak annual temperature fluctuations. This soil layer can reach a depth of 10 or 20 m (Fitzpatrick 1980, p. 26).

In an irrigation experiment (Vahrson et al 1986; Blume et al. 1985) at Bir Tarfawi, lower temperature fluctuations were measured in moist than in dry sandy and loamy soils. However, the depth of the soil affected by the daily temperature fluctuations was greater in the wet than in the dry soil.

## 2.6 Wind

The wind blows mainly from the north, only in the winter it can come from west. According to the meteorological station of East Uweinat, the wind direction on the serir plain by Tarfawi is NE, N and occasionally NW. Other wind directions are rare.

The wind speed in the area of the Dakhla and Kharga Oases is between 4 (= 8 knots) and 20 m/s (= 39 knots) occasionally wind speeds of 40 m/s (= 78 knots) can occur (Beller Consult 1985). In this area storms with high wind speeds are often observed in spring. At the meteorological station in Tarfawi, the dominant wind speed is between 12 knots = 6.2 m/s and 22 knots = 11.3 m/s. They have the Beaufort force scale number 4–5 which are moderate breeze to fresh breeze (Fig. 2.7). About 20% of the year have wind speeds less than11 knots (0–3 = calm to gentle breeze) and 9% with more than 23 knots (6–7 = strong breeze to moderate gale). Wind speeds of more than 35 knots (8 = strong gale) are very rare on the Serir plain near Tarfawi.

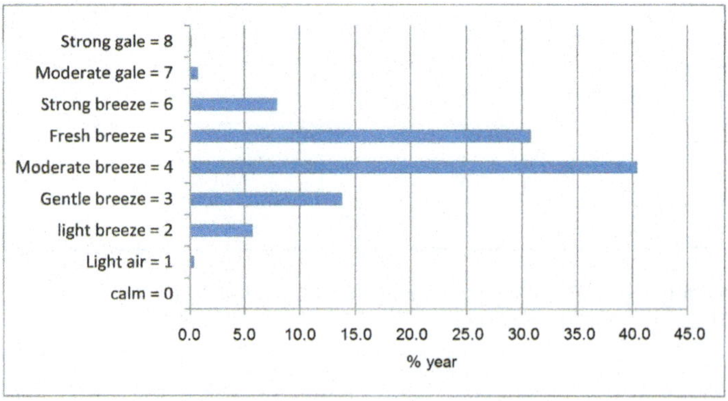


**Fig. 2.7** Wind speed distribution according to Beaufort wind force scale and according to the meteorological station of East Uweinat (Data from 1982 to 1986)

Within 5 years wind speed of 41 knots was only measured once by Tarfawi. Wind speeds of 25 knots (= 12.87 m/s) and more cause sandstorms and the migration and deformation of dunes (Photo 2.8). This wind speed accounts for 6.3% of the year (= 23 days a year). During sandstorms, the sand is blown and swirl a few meters in the air (Photo 2.9) while transported over relatively short distances in several stages. Sandstorms reshape dunes, set them in motion, and rub stones by sand blower that occur on the soil surface.

**Photo 2.8** Shape of dunes changes after every sandstorm

**Photo 2.9** During sandstorms, the sand is swirled only a few meters above the ground

**Photo 2.10** Huge clouds of dust from Egypt and Libya over the Mediterranean Sea moving toward Greece and Turkey. The island of Crete is almost completely covered (s. top left) (NASA 2003)

Compared to sandstorm, dust storm transports smaller particles at greater heights and over greater distances. The Khamassin is the name given to dust storms that occur in Egypt in early spring (mid-March to mid-May). They consist of warm air that blows from the south or southeast and is heavily polluted with dust. The air temperature during the Khamassin can be over 40 °C. The Khamassin is caused by a low atmospheric pressure developing south of the Mediterranean and moving to the east. The name Khamassin means "50" in Arabic because the storms occur about 50 days per year. The dust particles of the Khamassin can reach great heights and be transported up to over 1000 km. During the Khamassin storm, the air becomes hazy, and visibility can be severely reduced, so that in earlier years the airports and Harbors have to be closed. A satellite image from January 12, 2003, shows clouds of dust stretching from Egypt to Greece and Turkey (Photo 2.10). One dust storm can last three to four days continuously.

Most of the dust from the Khamassin and other dust storms comes from the full- and semi-arid regions of the Sahara (Photo 2.11). In the semi-arid region of northern Egypt, there are often short rain events during the Khamassin, which washes the dust from the air. Dust storms—as shown in Photo 2.11—were not observed in the investigated extreme arid region. This is because the land surface in the investigated area is mostly covered with sand and stones, wind storms can therefore blow only sand. In the extreme desert of southwest Egypt, dust storms do not affect the landscape, and the dust in the air is not deposited in situ, they only weaken the sunshine intensity in the area.

**Photo 2.11** Dust storm in the semi-arid southern Sahara on 18.10.1991 at Wad Medani in Sudan

# References


Beller Consult Ltd (1985) Water resources and soil potential development project New Valley. Freiburg/Siegen

Blume H-P, Vahrson W, Meshref H (1985) Dynamics of water, temperature and salts in typical aridic soils after artificial rainstorms. Catena 12:343–362

Briem E (1977) Beiträge zur Genese und Morphodynamik des ariden Formensatzes unter besonderer Berücksichtigung des Problems der Flächenbildung. Berliner Geograph. Abh 26:89

Fitzpatrick EA (1980) Soils; their formation, classification and distribution, pp 353. London and New York

Flohn H (1982) Warum ist die Sahara trocken? In: Mensching (Hrsg) Physische Geographie der Trockengebiete, Darmstadt, pp 55–69

Jung C (2007) Rain in Ancient Egypt, a linguistic approach. In: Bollig M, Bubenzer O, Vogelsang R, Wotzka H-P (eds) Aridity change and conflict in Africa. Proceedings of an international ACACIA conference held at Königswinter, Oktober 2003, Heinrich-Barth-Institut, pp 331–344, Köln

Meckelein W (1959) Forschung in der Zentralen Sahara, p 181, Braunschweig

NASA (2003) A catalog of NASA images and animations of our home planet. Visibleearth.NASA.gov

Vahrson WG, Blume H-P, Alaily F ( 1986) Simulation of episodic rainfalls in the western desert of Egypt. Berliner geowiss. Abh. (A) 72:55–67

# Chapter 3
# Hydrology

**Abstract** The Dakhla Basin belongs to the Nubian Aquifer System. The thickness of the aquifer in the Dakhla area is between 900 and 1600 m, in the Kharga area 300–800 m and at Tarfawi at about 500 m. Most of the groundwater of the aquifer was enriched in pluvial times. Today, groundwater in southwestern Egypt is considered fossil and nonrenewable. The groundwater in the oases is confined and is found on two layers at 100–350 m and 400–2000 m depth. In prehistory, the water must have flowed from several artesian wells. Due to water consumption in the last 5000 years, the groundwater level dropped for several meters. Due to the lowering of groundwater, many arable fields are watered sparingly, which is leading to soil salinization. According to the United States Salinity Laboratory Staff (United States Salinity Laboratory in Diagnosis and improvement of saline and alkali soils. US Department Agriculture Handbook, 1954), the quality of groundwater of the aquifer is classified as "medium" to "highly saline." Through millennia of irrigation in the area of the oases, a higher groundwater layer and small lakes (evaporation ponds) have formed.

**Keywords** Dakhla Basin · Nubian Aquifer · Evaporation ponds

## 3.1 The Aquifer

The Dakhla Basin (SW Egypt) and Kufra Basin (SO-Libya and NO-Chad) together form the Nubian Aquifer System, one of the largest groundwater reservoirs in North Africa. This aquifer consists mainly of continental sediments from Cambrian to Turonian age. The thickness of the Aquifer northwest of Farafra and at Kufra (Libya) is 3500 m. The effective porosity of sediments of the Nubian Aquifer System is given by Heinl and Thorweihe (1993) with 7–10%. This leads to a groundwater mass of 150,000 km$^3$, which would correspond to 1700 years of Nile water runoff, or a lake of 80 m depth over the entire aquifer. This corresponds to a lake with an area larger than that of the Federal Republic of Germany.

The basin is bordered to the east, south, and southwest by crystalline rocks and to the north and northwest by salty groundwater (Heinl and Thorweihe 1993). The



F. Alaily, *Ecosystems of Extremes*, Advances in Geographical and Environmental Sciences, https://doi.org/10.1007/978-3-031-76910-8_3

area studied in southwestern Egypt belongs to the southern part of the Dakhla Basin. Figure 3.1 shows the thickness of the aquifer in southwestern Egypt (Hesse et al. 1987). The thickness of the aquifer in the Dakhla area is between 900 and 1600 m, in the Kharga area 300–800 m and at Tarfawi at about 500 m (Müller 1981; Heinl and Thorweihe 1993; Thorweihe 1990).

The depth of groundwater generally decreases from Gilf Kebir (in southwest) toward Bir Safsaf (in southeast) and toward Dakhla and Kharga (in northeast). In the area of Gilf Kebir, the groundwater level is about 300 m deep; at Bir Tarfawi, it lies at a depth between zero and a few decimeters. Through millennia of irrigation in the area of the oases, a higher groundwater layer and small lakes (evaporation ponds) have formed. The secondary groundwater layer is located at a depth of 0.9–1.3 m. Often this groundwater disappears a few weeks after stopping irrigation.

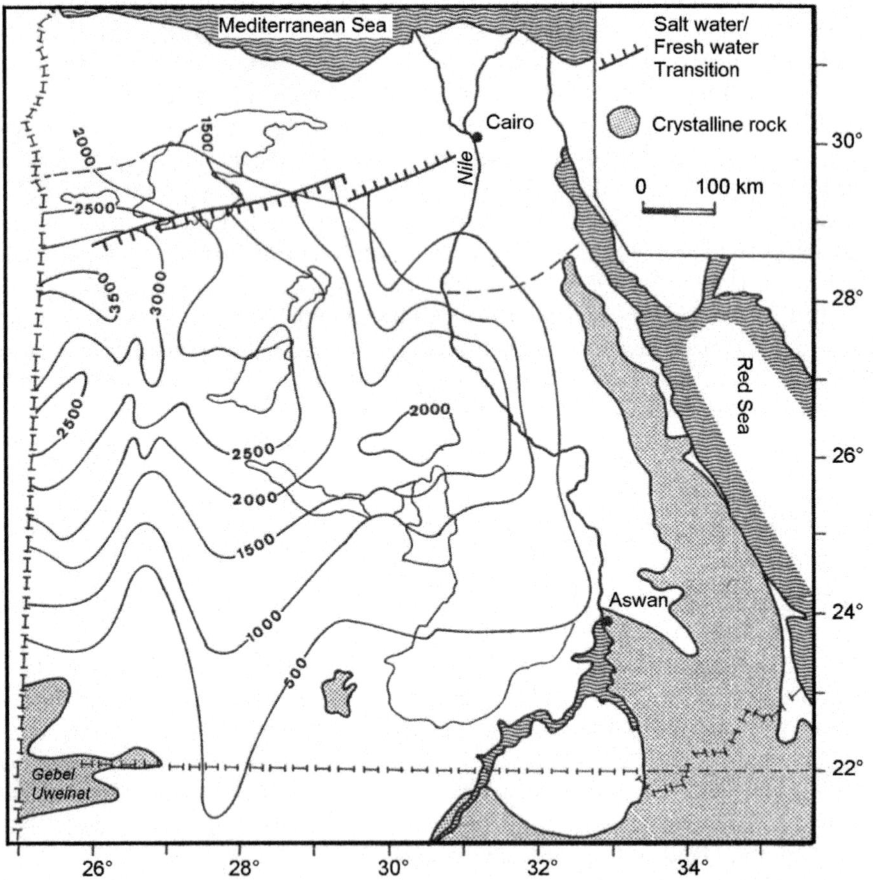


**Fig. 3.1** The aquifer thickness in meters (Hesse et al. 1987)

Most of the groundwater of the aquifer was enriched in pluvial times. In particular, between 34,000 and 28,000 ka and between 24,000 and 20,000 years ago, groundwater accumulation took place (Thorweihe 1986). During this time, the climate must have been humid. Stratigraphic studies of playa sediments suggest that between 9000 and 5800 years ago, three humid phases occurred in the Holocene, separated by arid intervals (Haynes 1980). The period from 22,000 to 11,000 years ago is considered as a hyper-arid phase. Today, groundwater in southwestern Egypt is considered fossil and nonrenewable. Witnesses of humid phases in the Central Sahara are archeological finds (hand axes: Photo 3.1, grinding stones: Photo 3.2, binding stones: Photo 3.3). Photo 3.4 presents a stone found on the Gilf Kebir Plateau which looks like a sculpture of a human head with African features. Also finds of animal remains that live in the savannah today, such as ostrich eggs (Photo 3.5), giraffe bones (Pachur and Röper 1984) and (Photo 3.6), and remains of termite nest are witnesses for past humid phases.

The groundwater in the oases is confined and is found on two layers at 100–350 m and 400–2000 m depth. In prehistory, the water must have flowed from several artesian wells.

Due to water consumption in the last 5000 years, the groundwater level dropped for several meters. Since Roman times until the beginning of the fifties, the groundwater level at the Kharga Oases dropped by 10 m (Murray 1952) (Photo 3.7). Since 1959, groundwater lowering has increased due to intensive water consumption as a result of the increased number of wells and the use of diesel and electric pumps. A subsequent delivery of groundwater from the huge Nubian Aquifer is extremely slow. The average regional permeability of the upper 1000 m of the Nubian rock formations in the oasis region is about 3.5–4.5 m per day, so that nonrenewable groundwater will continue to flow for a longer period of time (Müller 1981). Heinl and Thorweihe (1993) estimate

**Photo 3.1** Hand axes are found in different locations around playas

**Photo 3.2** A grinding bowl, for rubbing grains

**Photo 3.3** Binding stones with lace-up trace: cattle may have been tied to such stones

that groundwater is moving toward the depression at an average speed of a few meters per year. This is very slow compared to the consumption rate. This means that the water consumption in the oases is greater than the groundwater inflow from the water reservoir. As a result, the piezometric groundwater level sinks in the wells almost everywhere, and the old shallow wells gradually dry up. Therefore, more deep wells are drilled (Fig. 3.2). Due to the lowering of groundwater, many arable fields are watered sparingly, which is leading to soil salinization (Photos 3.8 and 3.9).

**Photo 3.4** This stone was found on the Gilf Kebir Plateau which looks like a sculpture of a human head with African features

**Photo 3.5** Fragments of ostrich eggs found on a Serir plain west of Black Hill

According to the United States Salinity Laboratory Staff (1954), the quality of groundwater of the aquifer is classified as "medium" to "highly saline." The Sodium Adsorption Ratio (SAR) has a relatively wide spectrum (see chap. groundwater quality).

**Photo 3.6** Giraffe bones embedded in playa sediment was found 1981 by Pachur (left) and Altmann (center) in Wadi Bakht in Gilf Kebir

**Photo 3.7** Dried water well (SW of the oasis Bulaq), which was an artesian spring in Roman times

## 3.2 Surface Water

In the vicinity of the investigated region lays "Lake Nasser" one of the largest water reservoirs in the world. It was named after the Egyptian President Nasser and was filled after the completion of the Aswan High Dam in 1970. The reservoir stretches

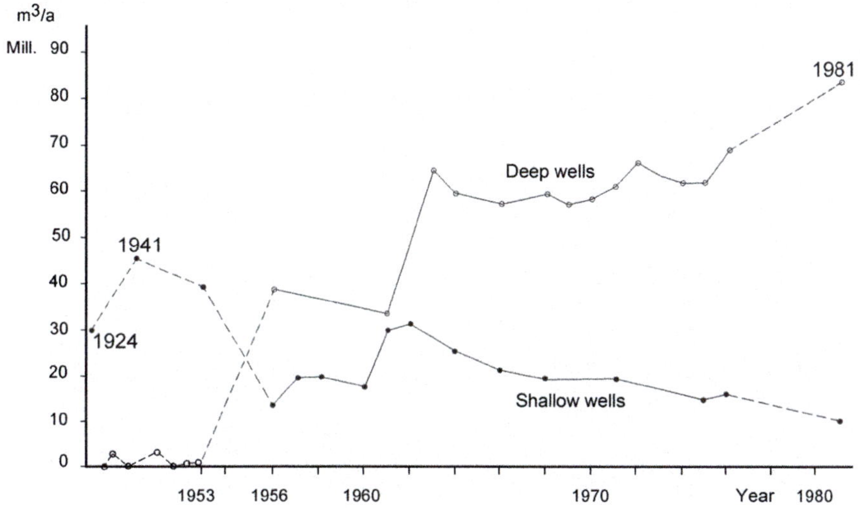


**Fig. 3.2** Development of water extraction from shallow wells and deep wells in the Kharga Oases from 1924 to 1981 (according to Beller Consult 1985)

**Photo 3.8** Abandoned fields north of Kharga

from Aswan to Sudan and has a length of about 550 km and an average width of 20 km. Its capacity varies between 135 and 169 million m$^3$. The surface of Lake Nasser varies between 5250 and 5860 km$^2$, depending on precipitation in the source area of the Nile. The water level varies between 151 and 183 m above sea level. Thanks to the huge amount of stored water, the lake provided Egypt with water during the 10 dry years (1970–1980) in North Africa.

**Photo 3.9** Abandoned fields heavily enriched with salts (east of the oasis Bulaq)

From 1998 to 2001, excess water in Lake Nasser created three large and 2 small lakes in the Toshka depression (Photo 3.10). The first lake was built in 1998 and is located 50 km west of Lake Nasser. By late 1999, three additional lakes formed successively westward, and the westernmost lake started forming sometime between September 2000 and March 2001. The five lakes held about 19.8 trillion cubic meters (= 700 trillion cubic feet) of water (NASA 2003) and had a surface area of 1500 km$^2$. The amount of water accumulated in the Toshka Lakes was about a quarter of the annual Nile water flowing to Egypt. Since 2000, rainfall in source areas of the Nile has decreased, so that the water supply to the Toshka Lakes has been interrupted since then. Due to evaporation and infiltration, the water level in the lakes decreases continuously. On satellite images from 2012 and 2020, only two lakes can be seen, three lakes dried out.

At present, several wells are being drilled for agricultural us on the Selima Serir in the southern part of the investigated region. Satellite images of Landsat 8 from February 26, 2017, showed large number of circular agricultural fields. Each circular has a diameter of 1 km and has a deep well in the center, which supplies a huge sprinkler system with groundwater.

In 2001, the Egyptian government began digging a canal that should bring water from Lake Nasser to the oases Baris and Kharga. A pumping station was built on Lake Nasser to supply the canal with water. A large part of the canal has been dug, and in 2002, the pumping of water from the lake into the canal began. Satellite images (www.Visibleearth.nasa.gov) from 2005 show the main canal with secondary canals that bring the water to the fields.

The Toshka Lakes and both agricultural projects (irrigation with groundwater and irrigation with Nile water) will have a noticeable impact on the groundwater conditions in SW Egypt.

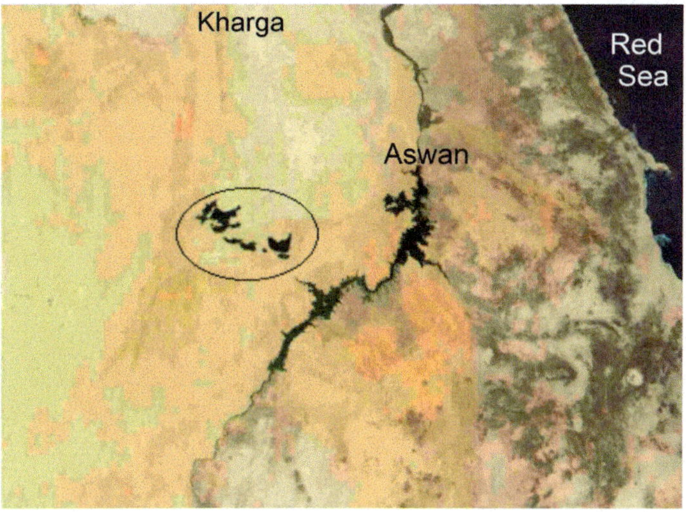


**Photo 3.10** Satellite image from August 2002 showing southern Egypt with the Toshka Lakes (circled) and Lake Nasser south of Aswan (NASA 2003)

## References

Beller Consult Ltd (1985) Water resources and soil potential development project New Valley. Freiburg/Siegen

Haynes CV (1980) Geological evidence of pluvial climates in the Nabta Area of the Western Desert, Egypt. In: Wendorf F, Schild R (eds) The prehistory of the Eastern Sahara. Academic press, New York, pp 355–372

Heinl M, Thorweihe U (1993) Groundwater resources and management in SW Egypt. In: Meisner B, Wycisk W (eds) Geopotential and ecology—analysis of a desert region. Catena Supplement 26, Cremlingen, pp 99–121

Hesse K-H, Hissene A, Kheir O, Schnäcker E, Schneider M, Thorweihe U (1987) Hydrogeological investigations in the Nubian Aquifer system eastern Sahara.- Berliner geowiss. Abh. (A) 75(2):397–464

Müller H-D (1981) Die Entwicklung des Oasengebietes von al-Kharga (Ägypten) als Resultat ökologischer Bedingungen und sozioökonomischen Wandels. Geographische Diplomarbeit, Göttingen

Murray GW (1952) The water beneath the Egyptian Western Desert. Geogr J 118:443–452

NASA (2003) A catalog of NASA images and animations of our home planet. Visibleearth.NASA.gov

Pachur HJ, Röper P (1984) The Libyan western desert and northern Sudan during the late Pleistocene and Holocene. Berl Geow Abh A 50:249–285, Berlin

Thorweihe U (1986) Isotopic identification and mass balance of the Nubian Aquifer System in Egypt. Berliineer Geowiss. Abh. (A) 72:87–97

Thorweihe U (1990) Das Grundwasser der Ostsahara. Die Geowissenschaften 8:211–219

United States Salinity Laboratory (1954) Diagnosis and improvement of saline and alkali soils. US Department Agriculture Handbook 60, p 160

# Chapter 4
# Geology and Lithology

**Abstract** The Dakhla Basin consists of marine and continental sediments. Apart from the areas at Gebel Kamil and north of Bir Safsaf, which consists of granite and gneiss, the region consists mainly of Quaternary, Paleocene, Cretaceous, and Upper Jurassic sediments. The Quaternary sediments cover more than half of the land surface. The quaternary sediments consist mostly of alluvial sand sheets, aeolian sand sheets, dunes as well as playa sediments. In the wadis, there are fluvial sand and fluvial gravel deposits as well as gravel debris deposits at the foot of slopes from gullies. On the slopes of hills and scarps, debris and aeolian sand deposits are found. The tertiary is represented in the study region with five formations (Table 4.1). These are the Thebes, Esna, Garra, Tarawan, and Kurkur Formation. They occur in the north and east and are the only rock formations that contain limestones. Cretaceous sediments belong to the Nubian Sandstone series and consist of 6 formations. The Upper Cretaceous (Duwi, Mut, Maghrabi, and Taref Formation) consists mainly of sandy sediments, the Lower Cretaceous (Sabaya, Abu Ballas, and Six Hills Formation) of sandy to fine conglomerate as well as clay and silt sediments, which partly contain carbonate. Lower Cretaceous is found on the slopes of Abu Tartur, in the foreshore of Abu Tartur and south of Dakhla Oases to the Great Sand Sea. The Lower Cretaceous form a chain of hills that extend from southwest to northeast. On the Gilf Kebir Plateau, there are sandstones and siltstones, which belong to the Maghrabi and the Sabaya Formation. Both Formations are found on the Gilf Kebir Formation, which also consists of sandstones and clay stones. The Gilf Kebir Formation is divided into the Lower Cretaceous and Upper Jurassic. Its age is similar to that of the Abu Ballas and Six Hills Formation. The Gilf Kebir Formation covers large areas south of the Gilf Kebir and an area near Gebel Kamils. The largest area of granite and gneiss is located north of Bir Safsaf. Small granite and Quatzitic sandstone outcrops can be found west of Gebel Kamil and near Nusab El Balgum. Basalt can be found on small areas mainly in the southwest and on the plateau of Gilf Kebir.

The original version of the chapter has been revised. A correction to this chapter can be found at https://doi.org/10.1007/978-3-031-76910-8_15



F. Alaily, *Ecosystems of Extremes*, Advances in Geographical and Environmental Sciences, https://doi.org/10.1007/978-3-031-76910-8_4

**Keywords** Thebes formation · Esna formation · Garra formation · Tarawan formation · Kurkur formation · Duwi formation · Mut formation · Maghrabi formation · Taref formation · Sabaya formation · Abu Ballas formation · Six Hills formation · Gilf Kebir formation · Granite · Gneiss · Basalt · Quartzite sandstones

The studied region is located in the Dakhla Basin. This basin and the Upper Nile Basin to the east are bounded by the Pan-African Orogen Zone of the Eastern Desert in the east, by an Uplift system that runs from the Uweinat area via Bir Safsaf to Aswan, to the south, and by the Gilf Kebir Uplift in the west. The Dakhla and the

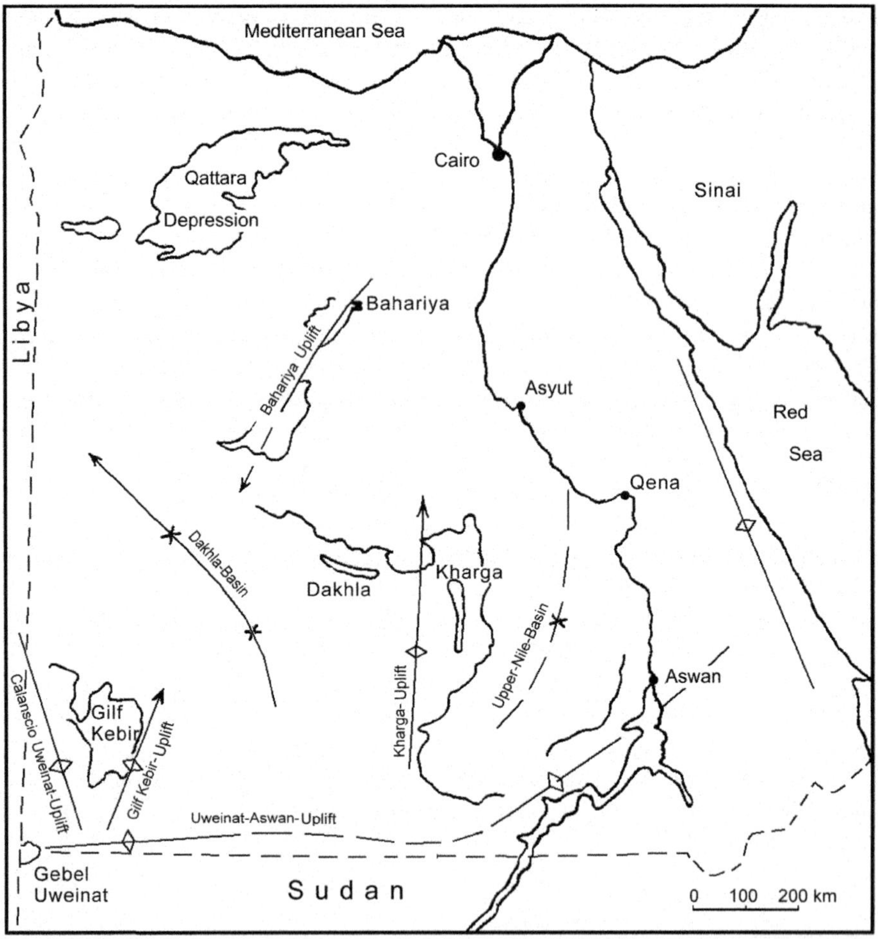


**Fig. 4.1** Tectonic main elements in the region of the Dakhla and the Upper Nile basin (according to Bisewski 1982)

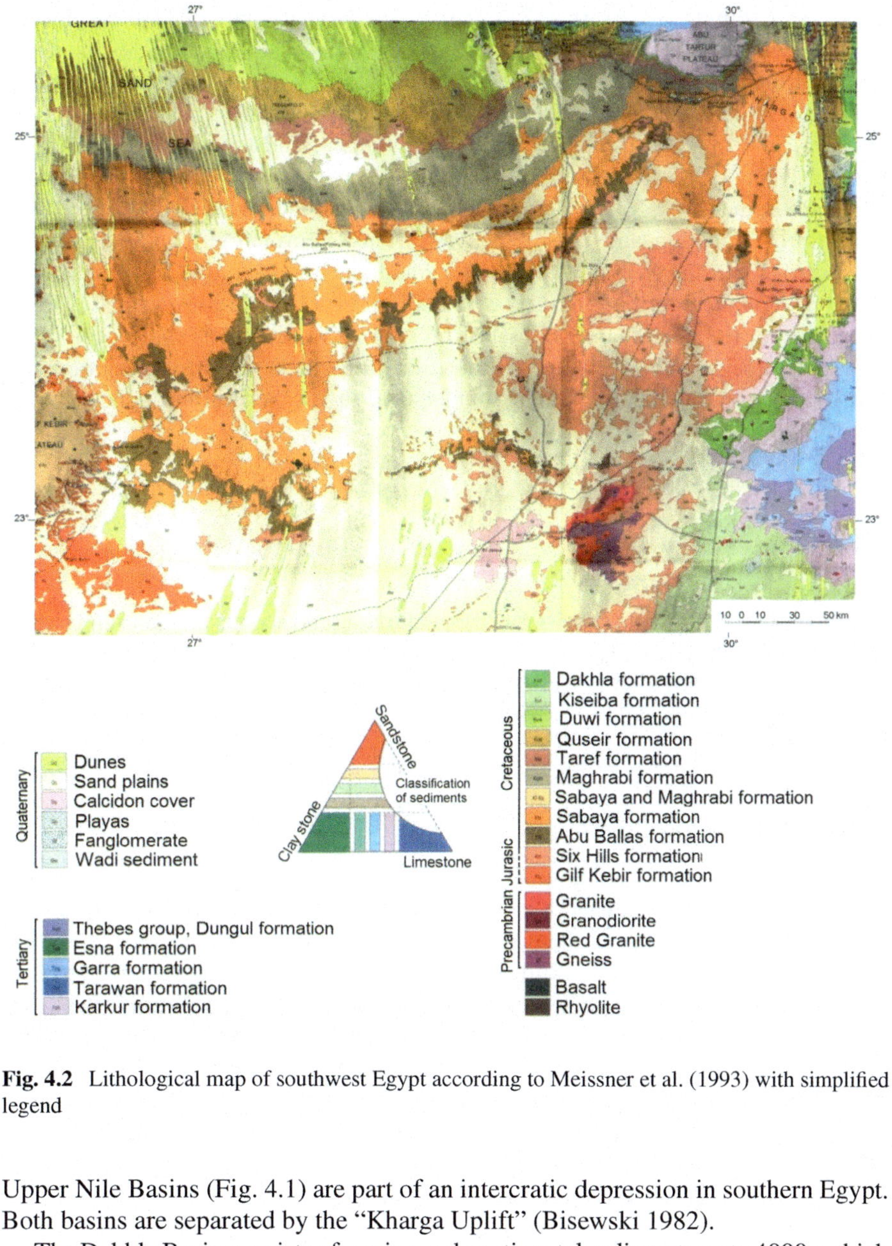


**Fig. 4.2** Lithological map of southwest Egypt according to Meissner et al. (1993) with simplified legend

Upper Nile Basins (Fig. 4.1) are part of an intercratic depression in southern Egypt. Both basins are separated by the "Kharga Uplift" (Bisewski 1982).

The Dakhla Basin consists of marine and continental sediments up to 4000 m thick (Said 1962; Hendriks et al. 1987). Apart from the areas at Gebel Kamil and north of Bir Safsaf, which consists of granite and gneiss, the region consists mainly of Quaternary, Paleocene, Cretaceous and Upper Jurassic sediments (Fig. 4.2). The Quaternary sediments cover more than half of the land surface. The regional distribution of the various rock formations in the region is presented in the geological maps of Egypt 1:1,000,000 (Klitzsch and List 1980) and 1:20,000,000 (Egyptian Geological

Survey and Mining Authority 1981) and the lithological map 1:1,000,000 of SW Egypt (Meissner et al. 1993).

Under the quaternary and tertiary sediments lie the Mesozoic Formations, which in the older literature were named with the collective term Nubian Sandstone. In 1962, R. Said divided the upper part of the Nubian Sandstone into 6 formations. Klitzsch et al. (1979) divided the entire Nubian Sandstone into 8 formations: Dakhla Shale, Phosphate Bed, Variegated Shale, Taref Sandstone, Plant Beds, Desert Rose Beds, Baris Formation, and parts of the Lower Clastic Formation. This was extended by Hendriks et al. (1984) to 10 formations and partly renamed. The names of the different geological formations used by the different authors are shown in Table 4.1. Starting with the quaternary sediments, all rocks and sediments that occur in the investigated region are presented in the following chapters. The descriptions are focused on information necessary to understand the properties and genesis of the soils.

## 4.1 Quaternary

The quaternary sediments consist mostly of alluvial sand sheets, aeolian sand sheets, dunes as well as playa sediments. In the wadis, there are fluvial sand and fluvial gravel deposits as well as gravel debris deposits at the foot of slopes from gullies. On the slopes of hills and scarps debris and aeolian sand deposits are found.

### *4.1.1 Alluvial Sand Sheets*

The alluvial sand sheets have a sandy to fine gravely texture and are often thin-layered. They form large parts of the soil parent material of the Serir plains. The monotony of the Serir plains is often interrupted by quarzitic outcrops (Photo 4.1). Occasionally a thick cover of aeolian sand occurs over the alluvial sand sheet. The largest and most famous Serir is the "Selima Sand Sheet," which is located in the south of the investigated region and in North Sudan. The sediments of the "Selima Sand Sheet" were deposited mainly in Pleistocene (Haynes et al. 1993). The alluvial sands of other Serire were deposited in Pliocene and at the beginning of Pleistocene (Haynes 1981). On the Serir areas, the thickness of sediment (aeolian sand + alluvial sand) generally decreases from the south (> 3 m) to the north (< 1 m).

### *4.1.2 Dunes and Aeolian Sand Sheets*

Dunes and aeolian sand sheets occur in the area of the Great Sand Sea, west of the Kharga–Baris Oases and west to south west of Dakhla Oases. They consist mainly of fine to middle sand, which is often loosely and unconsolidated. Such areas can only

**Table 4.1** Designation from lithostratigraphic units occurring in southwest Egypt by several authors (Hendriks et al. 1984)

| | | | | Dakhla–Kharga | Gilf Kebir | Kharga–Baris | Darb El Arbain |
|---|---|---|---|---|---|---|---|
| | | | Said (1962) | Klitzsch (1978) Klitzsch et al (1979) | Klitzsch (1978) Klitzsch et al (1979) | Hendriks et al. (1984) | Hendriks et al. (1984) |
| | | | Formation | Formation | Formation | Formation | Formation |
| Old tertiary | Eocene | Lower | Thebes | Dakhla | | Thebes | Dungul |
| | Paleocene | Upper | Esna Shale | | | Esna | Garra |
| | | Middle | Chalk | | | Tarawan | Karkur |
| | | Lower | Dakhla | | | Dakhla | |
| Upper Cretaceous | | Maastricht | Phosphat | | | | Kiseiba |
| | | Campan | Variegated Shale | Duwi Variegated Shale | | Mut | |
| | | Santon | Nubia Sandstone | Taref Sandstone | | Taref | |
| | | Coniac | | | | | |
| | | Turon | | | | | |
| | | Cenoman | | Plant Beds | | Maghrabi | |
| Lower Cretaceous | | Alb | | Desert Rose Beds | Gilf Kebir | Sabaya | |
| | | Apt | | Lingula Shale | | Abu Ballas | |
| | | Barrème | | Basal Clastics | | Six Hills | |

**Photo 4.1** Serir plain of alluvial sand, outcroping quartzite sandstone in the foreground and dunes on the horizon

be visited with off-road vehicles and are difficult to drive. The thickest and largest sand sheet deposits are located in the area of the Great Sand Sea (Photo 4.2). There, the thickness of the accumulated sand sheet is 120–300 m (Haynes 1982). Also, the highest dunes and greatest density of dunes can be found in the area of the Great Sand Sea.

On the "Selima Sand Sheet," Barchans in several locations can be observed. Dunes could also be deposited in the wind protected sides of hills, escarpments, wadis, or plant shrubs (e.g., tamarisk bushes). In particular, southern slopes are often covered with aeolian sand.

Dunes that are perpendicular to the wadis are called barrier dunes because they block the flow of water in the wadis. They act like water dams (Photo 4.3) that hold back the water from episodic rains and thus allow some plants to grow. Such barrier dunes can be found in the west–east oriented wadis in Gilf Kebir, such as in Wadi Akhdar (branched off from Wadi Wassa), Wadi Bakht, and Wadi Maftuh. All three wadis are located in the southern and eastern part of the Gilf Kebir.

The color of the dunes is different in different locations. Those north and east of Gilf Kebir are reddish; south of Bir Safsaf and in the middle of the Selima Serir they are white, otherwise light yellow. These different colors indicate different mineralogical compositions. These differences depend largely on the color of the source material from where the sand was blown: they are reddish if they come from Abu Ballas rocks, yellow if they come from yellow sandstone, and white if they come from white alluvial sand or white limestone. The different colors of the dunes can also be due to different phases of soil formation—in situ or allochthonous. The red

**Photo 4.2** Dunes from one of the largest Ergs in the Sahara, known as the Great Sand Sea

**Photo 4.3** Barrier dunes deposited across wadis act like water dams during episodic rain events

**Photo 4.4** Stone-covered relic dune in Wadi Maftuh in Gilf Kebir

dunes are relict and may have been developed by intensive weathering during earlier humid and warm periods. Such relic dunes can often be seen in the wadis of Gilf Kebir, where they are mostly covered with stones (Photo 4.4).

With the exception of the playas, a thin layer of aeolian sand (1–30 cm) is found in almost the entire region. During a sandstorm, parallel sand stripes were observed in a Serir area south of the Abu Ballas plateau (Photo 4.5). During the storm, the wind blew along the stripes and with every gust of wind, the stripes moved in stages in the direction of the wind.

### *4.1.3 Wadi Terraces*

In the recent water gullies in the upper part of the wadis, fine stratified loose sand is usually deposited, while in the lower parts, dense and fine stratified silty sand is accumulated and in the end pans of wadis playa sediments. In some wadis of Gilf Kebir and especially of the Egyptian plateau, wadi terraces (gravel terraces) are found (Photo 4.6). Most terraces are made of gravely fanglomerates with poorly sorted and poorly rounded material. Almost all terraces were built in Pleistocene during pluvial times. Using artifacts Caton-Thompson (1952) distinguished 3–4 terraces in wadis of the escarpment east of the oasis Bulaq. The formation of the higher terraces (25/35 m and 16/20 m) is assumed for the late tertiary and the beginning of the Pleistocene, the 5 m and 7 m terrace for the middle Pleistocene and the youngest terrace (3 m) for the late Palaeolithic. In the area of Gilf Kebir, wadi terraces are usually not

**Photo 4.5** Strips of aeolian sand on a Serir plain south of Abu Ballas Plateau

available. An exception is Wadi Maftuh, which has a terrace 100 m wide, 3 km long and which is 10–15 m above the Wadi bed (Kröpelin 1989). Wadi terraces in the area of Abu Ballas escarpment have not been observed. It seems that the aquatic terrain morphology there has been blurred by wind erosion.

### *4.1.4 Playa Sediments*

Playa sediments are clayey and loamy mud sediments. They are sediments from ephemeral lakes build during wet periods of Holocene. At that time, the surface water from artesian wells of oases and streams of water from escarpments could transport large amounts of silt and clay into the valleys (wadis) and deposit them in the lowest part of the landscape as playa sediments. They are found in depressions, in foreland of the Egyptian plateau and Abu Ballas plateau. In the Gilf Kebir, they are found in wide wadis and in end pans of some wadis (Photo 4.7). In addition, playa sediments have been deposited behind barrier dunes in some wadis of the Gilf Kebir. Such playa is found in wadi Akhdar (a branch of Wadi Wassa) and wadi Bakht. The playa of wadi Akhdar is 2 × 1.5 km large and 8 m deep (Kuper 1981). Measurements of playas near Abu Ballas revealed thicknesses of more than 18 m (Vahrson 1980).

The color of playa sediments seems to depend on their parent material: In the depressions of the Dakhla Oases and Kharga–Baris, they are predominantly brown to red, at Abu Ballas red and in the foreland of basalt cones gray. They are often

**Photo 4.6** Fluvial terraces are mainly found in wadis and in foreland of the north and east plateaus. The picture shows a wadi terrace in the foreshore of Abu Tartur Plateau

**Photo 4.7** Loamy lake sediment known as playa sediment in the foreland of Abu Ballas Plateau

stratified with sandy layers of different thicknesses. Clay-rich playa sediments are rarely covered by aeolian sand due to their smooth surface. Only in the area of the Great Sand Sea, playas are partly covered with aeolian sand or dunes.

### 4.1.5 Gravel and Debris Covers

Especially in the eastern foreland of the Gilf Kebir escarpment as well as in the southern foreland of the Egyptian escarpment, alluvial fans (flood gravel) can be found. These consist of poorly sorted fanglomerates, which must have originated during the pluvial period. Their accumulation was due to the abrupt reduction of surface water speed, so that most of the water cargo remained unsorted. This happens where the slope suddenly subsides, or the water-carrying channel suddenly becomes wide.

The slopes and the immediate foreland of escarpments and various hills, which consist of tertiary and Mesozoic rocks, are largely covered with mighty debris and/or aeolian sand (Photos 4.8 and 4.9). The debris is mostly conical. At the foot of escarpments, several debris cones are often intertwined and form long-stretched chains of debris cones, also known as the "Schuttfuß" (Leser et al 1984). Deposits of landslide were observed along almost the entire slopes of Abu Ballas escarpment (Kröpelin 1993).

## 4.2 Tertiary

The tertiary is represented in the study region with five formations (Table 4.1). These are the Thebes, Esna, Garra, Tarawan, and Kurkur Formation. They occur in the north and east and are the only rock formations that contain limestones.

The *Thebes Formation* consists of solid banked limestone with intermediate flint bands (Said 1962), which have an early Miocene age. The limestone is found on the surface of the plateau east of the oasis Bulaq and parts of the northern plateau. Due to erosion, the thickness of the Thebes Formation is only 55–60 m (Luger 1985). The *Esna Formation* is located under the Thebes Formation and consists of old Paleocene thin-layered green-gray clay and gray marl (Photo 4.10). Gypsum bands and iron oxide stains are spread out all over the thin-layered clay. They occur at the eastern escarpment (east of the Kharga–Baris Oases). East of Bulaq, its thickness is 80–180 m (Said 1962; Smettan 1987). From the Kharga Oases to the north, its thickness decreases.

On the northern and southeastern plateaus, rocks of the *Garra Formation* can be found. They consist of dolomite, limestone, marl, and clay stone, whose thickness is 25 m east of Baris. In addition to the Garra, the *Tarawan Formation* can be found on the plateau of the northern and eastern plateaus. It consists of compact limestones that contain marl layers. Characteristic of the limestone is a snow-white color (Photo

**Photo 4.8** Several generations of debris on slopes of hills and scarps

4.11). At its base are marl conglomerates (Photo 4.12), which occasionally contain phosphate concretions (Hendriks et al. 1984). East of Kharga–Baris, the Tarawan Formation has a thickness of 4–20 m, toward the south it gradually passes into the Garra Formation.

The *Kurkur Formation* occurs on the plateau of Abu Tartur and the southeast plateau. It consists mainly of claystones, siltstones, sandstones, and conglomerate (Hendriks et al.1987). Its thickness is 5 m east of Bulaq and 120 m near Abu el-Husein in the southeast. The rocks of the Kurkur Formation have been deposited mainly in the Old and Middle Paleocene.

In the Paleocene, marl stones containing phosphate were deposited under marine conditions. They form the *Dakhla Formation*, which occurs on small areas in the studied region. It can be found on the slopes of the Abu Tartur Plateau and in the northern part of the investigated area as well as in southeast and in the Great Sand Sea. On the lithological map of Meissner et al. (1993) this formation is put to Upper Cretaceous.

**Photo 4.9** Debris from rock blocks in the upper reaches from a side wadi of Wadi Dayek east of Gilf Kebir

**Photo 4.10** Green-gray clay of the Esna Formation east of the oasis Bulaq

**Photo 4.11** The Tarawan Formation on the northern and eastern plateau consists of compact snow-white limestones

## 4.3 Upper Cretaceous

Cretaceous sediments belong to the Nubian Sandstone series and consist of 6 formations (Table 4.1). The Lower Cretaceous consists mainly of sandy sediments, the Upper Cretaceous of sandy to fine conglomerate as well as clay and silt sediments, which partly contain carbonate.

In Upper Cretaceous, the *Duwi Formation* was formed under marine conditions. It consists of phosphate deposits as well as black silt stones and sandstone containing glauconite (Photo 4.13). The middle and upper parts of this formation contain different proportions of carbonate. It occurs on the slopes of Abu Tartur Plateau and on a large area in the north part of the studied area.

In Upper Cretaceous, the *Mut Formation* (or Quseir Formation in the lithological map of Meissner et al. (1993) can be found. It consists of claystones, siltstones, and sandstones. Between Kharga and Baris, the Mut Formation has a thickness of 60–80 m. It consists of gray, purple, and green claystones (Photo 4.14) which are sandy to silty as well as gray-white sandstone. These sediments have been deposited largely in marine and under marsh-like conditions (Hendriks et al. 1987). They can be found south of areas with Duwi Formation, and in the foreshore of Abu Tartur Plateau and its slope foot as well as in the Kharga–Baris Depression.

The Maghrabi and Taref Formations belong to Middle Cretaceous. The *Maghrabi Formation* was formerly called Plant Beds (Klitzsch 1978). It is located around the Kharga Oases and in the foreshore of Abu Tatur Plateau and it stretches from there

**Photo 4.12** Marly conglomerates can be found at the base of the Tarawan Formation

to the Great Sand Sea in the west. It consists largely of sandstone, siltstone, and claystone. The thickness of this formation is between 7 and 200 m (Klitzsch 1978; Hendriks et al.1987). The *Taref Formation* lies under the Maghrabi Formation and consists mainly of continental sandstones. It can be found around the Kharga Oases, at the foot of the Abu Tartur escarpment and in a narrow area that stretches south of the Dakhla Oases to the Great Sand Sea. At Dakhla, the Taref Formation reaches a thickness of 110 m.

## 4.4 Lower Cretaceous

The *Sabaya Formation* by Klitzsch (1978) referred to as Desert Rose Beds belongs to the Lower Cretaceous. It consists of continental deposits characterized by sandstones and rust-stained paleosoil (Photos 4.15 and 4.16). Occasionally, sandy conglomerates are found (Hendriks et al. 1987). The base of the Sabaya Formation consists of

**Phot 4.13** The Upper Cretaceous Duwi Formation consists of phosphate deposits as well as black siltstones (left in the center of the picture) and sandstone containing glauconite

**Phot 4.14** Gray-green clay stones of the Mut Formation in the east of Dakhla

white and kaolinite-rich sandstones. Petrified trees and fossils of plant remains are common in the Sabaya Formation. This formation covers large areas and can be found southwest of the Kharga Oases up to the Great Sand Sea. Its thickness is between 30 and 300 m (Bisewski 1982). At Gebel Qulu El Sabaya (25°21′ N and 29°43′ E) it is 160 m and it lies on the Abu Ballas formation.

Under the Sabaya Formation lies the *Abu Ballas Formation* and major part of the Six Hills Formation. The Abu Ballas Formation (named after Abu Ballas Hill 27°35′ E and 24°23′ N) or Lingula Shale Formation (Klitzsch 1978) consists mainly of solidified silt stones and clay stones (Photo 4.17). They have green to dark red colors and form a chain of hills that extend from southwest to northeast. The thickness of this formation is 40 m at Abu Ballas. It is found on the Six Hills Formation, which has an Upper Jurassic to Lower Cretaceous age. The *Six Hills Formation* consists largely of continental sandstone and fine gravel stones as well as conglomerates (Photos 4.18 and 4.19). The sediments of this formation lie over deeply weathered crystalline rocks (granite and gneiss). The rocks of the Six Hills Formation occur on large areas in the center and east of the investigated region (Meissner et al. 1993). Its thickness is several hundred meters. Within this formation, there are paleosoils as well as petrified trees (Photo 4.20).

**Photo 4.15** Continental sandstones of the Sabaya Formation

**Photo 4.16** The base of the Sabaya Formation consist of white sandstone

**Photo 4.17** The Abu Ballas Formation or Lingula Shale Formation (Klitzsch 1978) consists mainly of solidified green and red siltstones and clay stones

**Photo 4.18** The Six Hills Formation consists mostly of continental sandstone and fine gravel stones

**Photo 4.19** Similar to the Sabaya Formation, the Six Hills Formation consist partially of white sandstone

**Photo 4.20** Sandstones of the Six Hills Formation are rich in petrified trees

## 4.5 Lower Cretaceous and Upper Jurassic

On the Gilf Kebir Plateau, there are sandstones and siltstones (Photo 4.21), which belong to the Maghrabi and the Sabaya Formation. Both Formations belong to Lower Cretaceous and are found on the Gilf Kebir Formation (Meissner et al. 1993), which also consists of sandstones and clay stones. The Gilf Kebir Formation is divided into the Lower Cretaceous and Upper Jurassic (Klitzsch and Lejal-Nicol 1984). Its age is similar to that of the Abu Ballas and Six Hills Formation. The Gilf Kebir Formation covers large areas south of the Gilf Kebir and an area near Gebel Kamils.

## 4.6 Metamorphites and Magmatites

The largest area of granite and gneiss is located north of Bir Safsaf. Small granite and Quatzitic sandstone outcrops can be found west of Gebel Kamil (north of the Sudanese border) and near Nusab El Balgum. Occasionally, they occur within areas of the Six Hills Formation as well as in northeast of the Gilf Kebir (Photo 4.22). Granite can be recognized from a distance by their typical wool sack shaped monoliths (Photo 4.23).

Basalt can be found on small areas mainly in the southwest and on the plateau of Gilf Kebir (Photos 4.24 and 4.25). They usually form small isolated hills and "dykes" (Photo 4.26). The largest basalt surface is located north of Black Hill. Rhyolite (acid

**Photo 4.21** The Lower Cretaceous of the Gilf Kebir Formation consists mainly of brown sandstone and white siltstone

**Photo 4.22** The landscape northeast of Gilf Kebir is characterized by quartzite sandstone blocks

**Photo 4.23** Granite blocks west of Gebel Kamil

**Photo 4.24** Basalt columns polished by wind

volcanite) occurs only at Nusab El Balgum. Most of the volcanites in the studied region are originated in Tertiary.

**Photo 4.25** Top of a basalt hill in the central part of the studied region

**Photo 4.26** Basalt crevice (dam = dyke) southeast of Six Hills

## References


Bisewski H (1982) Zur Geologie des Dakhla-Beckens (Südwest-Ägypten). Sedimentologie und Geochemie der Nubischen Gruppe. Berliner geow. Abh. (A) 40:1–86

Caton-Thompson G (1952) Kharga Oases in prehistory. Athlone press, London

Egyptian Geological Survey and Mining Authority (1981) Geologic map of Egypt 1:2000000. Cairo

Haynes F (1981) The Darb El-Arbain desert: a product of Quaternary climatic change. Ann Geol Surv Egypt 11:91–117

Haynes F (1982) Great Sand Sea and Selima Sand Sheet, Eastern Sahara: geochronology of desertification. Science 217:627–633

Haynes CV, Maxwell TA, Johnson DL (1993) Stratigraphy, geochronology, and origin of the Selima Sand Sheet, eastern Sahara, Egypt and Sudan. In: Thorweihe, Schandelmeier (eds) Geosientific research in Northeast Africa. Balkema, Rotterdam, pp 621–626

Hendriks F, Luger P, Bowitz J, Kallenbach H (1987) Evolution of the depositional environments of SE-Egypt during the Cretaceous and Lower Tertiary. Berliner geow Abh (A) 75(1):49–82

Hendriks F, Luger P, Kallenbach, Schröder J-H (1984) Stratigraphical and sedimentological framework of the Kharga-Sin El-Kaddab stretch, Western Desert, Egypt. Berl Geow Abh A/50, 117–153

Klitzsch E (1978) Geologische Bearbeitung Südwest-Ägyptens. Geol Rundsch 67(2):509–520

Klitzsch E, Harms JC, Lejal-Nicol A, List FK (1979) Major subdivisions and depositional environments of Nubia strata, southwestern Egypt. AAPG Bull 63(6):967–974

Klitzsch E, Lejal-Nicol A (1984) Flora and fauna from strata in southern Egypt and northern Sudan. Berliner geowiss. Abh. (A) 50:47–79

Klitzsch E, List FK (1980) Egypt-geological interpretation map, 1:500000 (preliminary edn.). TFH Berlin

Kröpelin S (1989) Untersuchungen zum Sedimentationsmillieu von playas im Gilf Kebir (südwest-Ägypten). Africa Praehistorica 2:183–305

Kröpelin S (1993) Geomorphology, landscape evolution and paleoclimates of Southwest Egypt. Catena Supplement 26:31–65

Kuper R (1981) Untersuchungen zur Besiedlungsgeschichte der östlichen Sahara, Bericht über die Expedition 1980. Beiträge zur allgemeinen und vergleichenden Archäologie 3. Beck'sche Verlagsbuchhandlung, München, pp 215–275

Leser H, Haas H-D, Mosimann T, Paesler R (1984) Wörterbuch der Allgemeinen Geographie.- Bd 2, Deutscher Taschenbuch Verlag Westermann, Braunschweig, p 420

Luger P (1985) Stratigraphie der marinen Oberkreide und des Alttertiärs im südwestlichen Obernil-Becken (SW-Ägypten) unter besonderer Berücksichtigung derMikropaläontologie, Palökologie und Paläogeographie. Berliner geowiss Abh (A) 63:151

Meissner B, Ripke U, Wycisk P (1993) Lithology map, South-West Egypt 1:1000000. Catena Supplement 26

Said R (1962) The geology of Egypt. Elsevier, Amsterdam

Smettan U (1987) Typische Böden und Bodengesellschaften der Extremwüste Südwest-Ägyptens. Berliner geow Abh A 83:1–190

Vahrson WG (1980) Mächtigkeiten fossiler Stillwassersedimente in der östlichen Zentralsahara. Die Erde 111:140–146

# Chapter 5
# Landscape Development

**Abstract** The modeling of the land surface in the studied region began mid to end of tertiary. In the course of 60 million years of tertiary, tectonics and volcanism have created numerous basalt mounds and crevice, especially in the Gilf Kebir area and in center of the studied region. Under the humid climatic conditions of tertiary, intense and deep weathered soils were developed. Remains of iron crusts or hardened iron concretions of tertiary soils have been found in several locations on the soil surface due to their resistance against wind erosion. At the end of tertiary, the great wadis, such as those of the Gilf Kebir, Abu Ballas, and Abu Tartur, were formed by surface water and heavy land erosion. In addition, as a result of the progressive erosion of escarpments and plateaus, a number of inselbergs and "tafelbergs" were isolated. Such inselbergs are found today several hundred kilometers away from their escarpments. Karst phenomena (caves, dolines, remnants of stalactites, and stalagmites) are found on the limestone plateau near Abu Tartur as well as on the eastern escarpment and near Six Hills. The climate of the Sahara was dominated during Quaternary by arid conditions, which were interrupted by pluvial times. The transition to the hyper-arid climate in Eastern Sahara lasted more than 1500 years. It began approximately 4500 years b.p., so that only around 3000 years b.p. the Neolithic settlement came to an end. The backward movement from escarpments of plateaus took place not only with the help of surface water erosion and landsliding in humid climates, but also during the dry periods of Quaternary due to fall wind and corrosion. Corrosion and deflation caused the formation of the Dakhla and Kharga depressions. Deflation plains were developed under arid climate conditions and the absence of vegetation cover. They are mainly formed by erosion rather than sedimentation. Deflation plains consist mainly of Mesozoic rocks, granite, or alluvial sands. Deflation plains which are covered with dense and very coarse particles (middle gravel to stones) are called hamada. In comparison, Serirs are covered by residual of fine gravel and coarse sand.

**Keywords** Tertiary · Quaternary · Pluvial time · Hyper-arid · Neolithic

The modeling of the land surface in the studied region began mid to end of tertiary (Kröpelin 1993). In the course of 60 million years of tertiary, tectonics and volcanism

 
F. Alaily, *Ecosystems of Extremes*, Advances in Geographical and Environmental Sciences, https://doi.org/10.1007/978-3-031-76910-8_5

have created numerous basalt mounds and crevice, especially in the Gilf Kebir area and in center of the studied region. However, the actual land surface is characterized less by tectonics than by exogenous forces. Under the humid climatic conditions of tertiary, intense and deep weathered soils were developed. Traces of such paleosoils can still be found in the landscape. Remains of iron crusts or hardened iron concretions of tertiary soils have been found in several locations on the soil surface due to their resistance against wind erosion (Photo 5.1). More over a well-developed and strongly solidified plinthite horizon of a Plinthosol (formerly laterite) was found in Wadi Wassa in the south of Gilf Kebir (Photo 5.2). At the end of tertiary, the great *wadis*, such as those of the Gilf Kebir, Abu Ballas, and Abu Tartur, were formed by surface water and heavy land erosion. In addition, as a result of the progressive erosion of escarpments and plateaus, a number of inselbergs and "tafelbergs" were isolated. Such inselbergs are found today several hundred kilometers away from their escarpments. Also the fluvial topography (called radar rivers) discovered by radar images under the aeolian sand of the Selima Sand Sheet is described as pre-quaternary (McCauley et al. 1982, 1986).

Karst phenomena (caves, dolines, remnants of stalactites, and stalagmites) can be found on the limestone plateau near Abu Tartur as well as on the eastern escarpment. Their formation is supposed to be in Pliocene under humid climate conditions (Kröpelin 1993).

Nearby (a few km) of the Six Hills, remains of stalactites and stalagmites were found within limestone debris of an inselberg. The inselberg is located between two basalt crevices.

The climate of the Sahara was dominated during Quaternary by arid conditions (Kröpelin 1993), which were interrupted by pluvial times. From stratigraphic studies and $^{14}C$ dating, the climate conditions in Eastern Sahara can be reconstructed as follows (Fig. 5.1). According to Pachur et al. (1987), the climate between 40,000 and 25,000 years b.p. in the entire region of Eastern Sahara was semiarid. During this period, groundwater was formed. From 25,000 to 11,000 years b.p., the investigated area was influenced by a hyper-arid climate with high aeolian activity (Pachur et al. 1987). According to Kröpelin (1993), the hyper-arid climate occurred before 9500 years b.p. Between 9500 and 6000 years b.p., the climate was arid and was characterized by heavy episodic precipitation (Kröpelin 1993). The climate change is documented by relatively rapid formation of lakes for the whole Eastern Sahara. Around 7000 years b.p., lakes formation reached their maximum extension (Pachur and Hoelzmann 1998). Apart from local freshwater lakes and swamps, drainage systems were also formed, which served as migration pathways for big mammals, so that they could spread over large areas of the Sahara (Pachur and Hoelzmann 1998). These animals served as hunting preys for the Neolithic population. The arid climate extends further from 6000 to 5000 years b.p. (Kröpelin 1993) without any significant special features. During this period, some winter rain may have occurred regularly with the west winds.

The transition to the renewed hyper-arid climate in Eastern Sahara lasted more than 1500 years. It began approximately 4500 years. b.p. (Pachur and Kröpelin 1989),

**Photo 5.1** Different forms of iron crusts and iron concretions. Remains from paleosoils

**Photo 5.2** Spotted and strongly hardened saprolith horizon of a tertiary soil

so that only around 3000 years b.p. the Neolithic settlement came to an end (Pachur and Hoelzmann 2000).

The maximum rainfall of the humid phases of early and middle Holocene was estimated by Kröpelin (1993) with 100 mm $a^{-1}$. Gabriel (1978) estimated the annual rainfall for the lowlands in southwest Egypt over the last 8500 years with less than 50 mm. Under these conditions, only a minimum of biological activity could take place. Some animals and plants have retreated to the Gilf Kebir, where the relief has induced a positive modification on the climatic conditions.

In the pluvial times, the wadis, which had been formed in Tertiary, were only slightly further modeled. Traces of precipitation of the pluvial period can often be observed as dry water runners at the foot of escarpments and hills (Photo 5.3). The surface water cargo at that time was deposited in foreland, where broad planes were formed. The "radar rivers" on the Selima Sand Sheet may have led water during the pluvial times (McHugh et al. 1988, 1989) and their sandy cargo contributed to the formation of wide plains.

During pluvial time of Pleistocene and the humid phases of Holocene, playas were deposited in wadi ends. Barrier dunes deposited across the wadis during arid times have functioned in the pluvial times like dams and water reservoirs. In the reservoirs, clay and loamy sediments have accumulated. Stratigraphic studies on a playa sediment in Wadi Akhdar (a Tributary of Wadi Wassa in the south of Gilf Kebir) showed a seven-time change of wet and dry phases covering a period from

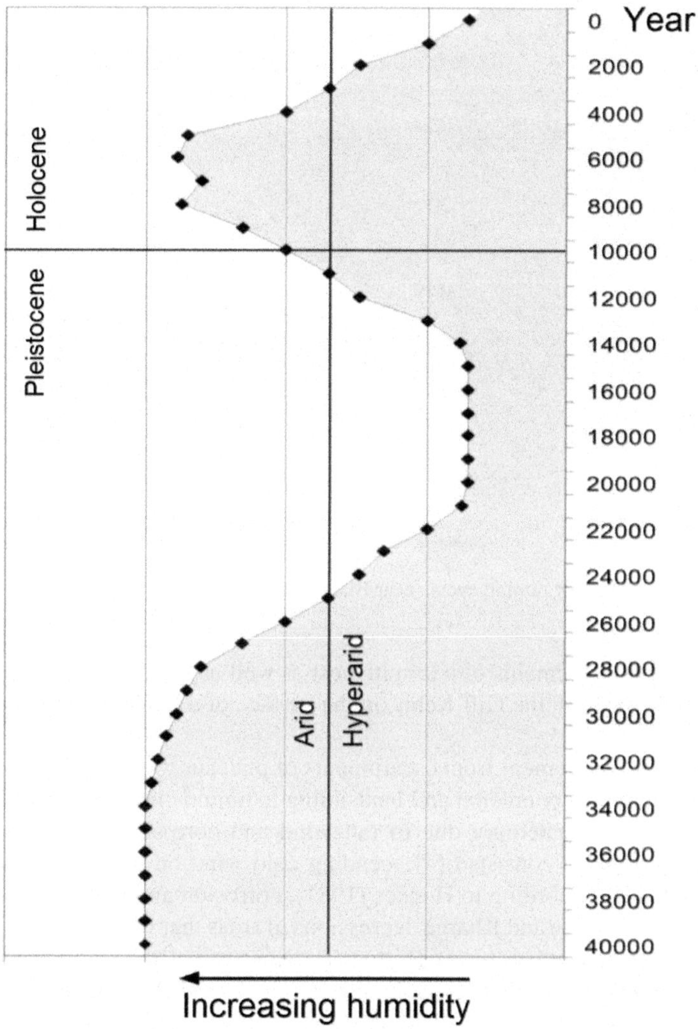


**Fig. 5.1** Simplified climate development in Eastern Sahara according to data from Pachur et al. (1987) and Kröpelin (1993)

the 8th to the 4th millennium BC (Kuper 1981). The average age of playa sediments is between 10,000 and 4000 years b.p. (Pachur and Röper 1984). The absence of any kind of molluscs or other aquatic fauna and flora is striking in all investigated playas. Today, the playa sediments are eroded by wind. The erosional remains are often formed as Yardangs (Photos 5.4 and 5.5). On the surface and around playas, there are often neolitic artifacts, such as grinding bowls, hand axes, etc. (Gabriel

**Photo 5.3** Traces of heavy rainfall events near Black Hill

1986; Kuper 1981). Remains of a termite nest as well as bones from a giraffe were found in Wadi Bakht in the Gilf Kebir on the surface of a playa (Pachur and Röper 1984).

The backward movement from escarpments of plateaus took place not only with the help of surface water erosion and landsliding in humid climates, but also during the dry periods of Quaternary due to fall wind and corrosion. Fall wind occurs usually at night and it consists of descending cold wind on lee side from escarpments of plateaus. According to Haynes (1981), corrosion and deflation caused the formation of the Dakhla and Kharga depressions in areas that were primarily favored by tectonics (ditches and hollows). Both depressions can also be called "Deflationswannen" (Deflated depressions) (Pfannenstiel 1953). The deflation process was caused by the descending winds. The erosion effect of the descending wind must have been stronger the greater the difference in altitude between plateaus and depressions. These differences in altitude increased rapidly: As soon as the land surface became groundwater influences and was enriched with salts; the salts accelerated the mechanical weathering of the rocks, so that the erosion forces of wind become more effective. This hypothesis is still controversial. According to Said (1983), Dakhla and Kharga depressions were formed by a hypothetical Pliocene river system.

Even larger than the Deflated depression ("Deflationswannen") are the Deflation plains, which also developed under arid climate conditions and the absence of vegetation cover. They are mainly formed by erosion rather than sedimentation. Deflation plains consist mainly of Mesozoic rocks, granite, or alluvial sands. On these landscapes, particles $< 600$ microns are blown by the wind, so that the land surfaces

**Photo 5.4** Yardangs from playa sediment in the foreland of Abu Ballas escarpment

**Phot 5.5** Southern side of the Yardangs of Abu Ballas with their aerodynamic shape

**Photo 5.6** Tri-edged windkanter

**Photo 5.7** Single-edged windkanter

are only covered with particles > 600 microns. Deflation plains which are covered with dense and very coarse particles (middle gravel to stones) are called hamada. In comparison, Serirs are covered by residual of fine gravel and coarse sand. On the surface of the Serirs and Hamadas, there are often "windkanters" with one to three edges (Photos 5.6 and 5.7). In addition, rocks found on the ground surface are usually polished by wind sand blowers (Photo 5.8).

**Photo 5.8** Sandstone polished by wind sand blower at Black Hill

The sand eroded by wind is deposited as aeolian sand sheet and as dunes, where the wind speed decreases as a result of a rough land surface (due to rocks or vegetation) or on the lee side of dunes, hills, escarpments, slopes of gullies, or vegetation islands. The source of the enormous amount of aeolian sands deposited in the Great Sand Sea as well as the alluvial sands of the "Selima Sand Sheet" is still unclear.

## References

Gabriel B (1978) Klima und Landschaftswandel der Sahara. In Museen der Stadt Köln: Sahara – 10000 Jahre zwischen Weide und Wüste, pp 22–34

Gabriel B (1986) Die östliche Libysche Wüste im Jungquartär. Berliner geogr. Szudien 19:1–214

Haynes F (1981) The Darb El-Arbain desert: a product of Quaternary climatic change. Ann Geol Surv Egypt 11:91–117

Kröpelin S (1993) Geomorphology, landscape evolution and paleoclimates of Southwest Egypt. Catena Supplement 26:31–65

Kuper R (1981) Untersuchungen zur Besiedlungsgeschichte der östlichen Sahara, Bericht über die Expedition 1980. Beiträge zur allgemeinen und vergleichenden Archäologie 3, Beck'sche Verlagsbuchhandlung, pp 215–275, München

McCauley JF, Schaber GG, Breed CS, Grolier MJ, Haynes CV, Issawi B, Elachi C, Blom R (1982) Subsurface Valeys and Geoarchaeology of the Eastern Sahara Revealed by Shuttle Radar. Science 218:1004–1020

McCauley JF, Breed CS, Schaber GG, McHugh WP, Issawi B, Haynes CV, Grolier MJ, El Kilani A (1986) Paleodrainages of the Eastern Sahara. The radar rivers revisited (SIR-A/B implications for a mid-tertiary trans-African drainage system). EEE Trans Geosci Rem Sens GE-24:624–648

McHugh WP, McCauley JF, Haynes CV, Breed CS, Schaber GG (1988) Paleorivers and geoarchaeology in the Southern Egyptian Sahara. Geoarchaeology 3:1–40

McHugh WP, Schaber GG, Breed CS, McCauley JF (1989) Neolithic adaption and the Holocene Functioning of Tertiary palaeodrainage in southern Egypt and northern Sudan. Antiquity 63:320–336

Pachur H-J, Hoelzmann P (1998) Zur Paläoökologie der Ostsahara. In: Böse M, Hofmann J (Hrsg) Berliner Geogr. Abh., 63, Berlin, pp 27–34

Pachur H-J, Hoelzmann P (2000) Late Quaternary palaeoecology and palaeoclimates of the eastern Sahara. J Afr Earth Sc 30:929–939

Pachur H, Kröpelin S (1989) L'aridification du Sahara oriental à l'Holocéne moyen et supérieur. Bull Soc Géol France 1989 (8e sér.) 5:99–107

Pachur HJ, Röper P (1984) The Libyan Western Desert and Northern Sudan during the Late Pleistocene and Holocene. Berl Geow Abh A 50:249–285

Pachur H-J, Röper H-P, Kröpelin S, Goschin M (1987) Late quaternary hydrography of the Eastern Sahara. In: Klitzsch E, Schrank E (Hrsg) Research in Egypt and Sudan, Reihe A/75.2, Berlin, pp 331–379

Pfannenstiel M (1953) Die Entstehung der ägypt. Oasendepressionen, das Quartär der Levante, II, Abh. Der Akademie der Wissenschaften, pp 337–411, Mainz

Said R (1983) Remarks on the origin of the landscape oft the eastern Sahara. J Afr Earth Sc 1:153–158

# Chapter 6
# Geomorphology

**Abstract** The Region is divided into 10 landscape units: Serirs, Ergs, Deflation plains, Deflated depression ("Deflationswannen"), hilly landscapes, and plateaus with escarpments, wadis, oases, and birs. *Serirs* are wide monotonous plains of alluvial sand and aeolian sand of different thicknesses. The surface of the Serir is covered with a layer of coarse sand to fine gravel. The monotony of the Serir is occasionally interrupted by dunes or by outcrop of quartzite sandstone or iron crust remnants. The famous Serirs in the studied region are: "Atmur El Kabeish" and the "Selima Sand Sheet". The *Ergs* are characterized by aeolian sand and dunes deposited on different substrates. Almost all types of dunes can be found in the investigated region: barchans (creeping dune or crescent dune), Seifs (longitudinal dunes), dome dunes, and transition forms. *Deflation areas* are landscapes formed by wind erosion. They are divided into Deflated plains and Deflated depressions (Deflationswannen). Deflated plains occupy large areas in southwestern Egypt. According to the main wind direction, the deflated depressions are mainly found south of the Egyptian and Abu Ballas escarpment. Its formation is largely due to the erosional effect of fall winds. Due to the selective erosion of wind, the ground surface of deflated areas is covered with a layer of stones and/or fine gravel and has a rougher micro-relief compared to Serir. Deflation plains occur on almost all tertiary and Mesozoic rocks. *Hilly landscapes* consist of "inselbergs" and "tafelbergs" from tertiary (in northeast) and Mesozoic rocks as well as basalt mounds (cones) and rarely quartzite sandstone mounds. Between the hills, playas, fluvial sediments, and aeolian sand are often found in different thicknesses. Small dunes or aeolian sand sheets can occur on the lee side of the hills. The most famous hills are Gebel Edmonstone, Six Hills, Abu Ballas, and Gebel Kamil. In the investigated region there are three *plateaus*: the Egyptian Plateau, the Abu Ballas plateau, and the Gilf Kebir Plateau. The Egyptian plateau includes the northern plateau with the Abu Tartur Plateau and the eastern plateau. The slopes of escarpments found at the edge of plateaus are covered with debris which leis over pediments of different widths. The edges of the plateaus and escarpments as well as the pediment areas are rugged and fissured by wadis and dry water channels. *Wadis* are found on the periphery of the Egyptian and the Gilf Kebir plateaus and extend to the foreland of their escarpments. Water flow in the wadis is approximately once every decade. Inhabited *oases* lie in two major depressions:

 
F. Alaily, *Ecosystems of Extremes*, Advances in Geographical and Environmental Sciences, https://doi.org/10.1007/978-3-031-76910-8_6

the Kharga-Baris and The Dakhla Depression. In these depressions, there are a large number of wells and settlements. There are several birs to the southeast of the study area. These are sites with high groundwater levels, often with vegetation and wildlife, but not inhabited by humans. The most famous birs are: Bir Tarfawi, Bir Safsaf, Bir Sahara, and Bir Misaha. The last two birs have no vegetation.

**Keywords** Atmur El Kabeish · Selima Sand Sheet · Barchan · Seif · Dome dune · Deflated plain · Deflationswanne · Deflated depression · Gilf Kebir plateau · Abu Tartur plateau · Egyptian plateau · Gebel Edmonstone · Six Hills · Abu Ballas · Gebel Kamil · Bir Tarfawi · Bir Safsaf · Bir Sahara · Bir Misaha

In order to make a soil map with a scale of 1:1,000,000 for southwest Egypt, the region was divided into smaller units. Each unit had to be as homogeneous as possible in terms of relief, groundwater depth, and soil parent material. According to the topographical maps of SW Egypt (Meissner and Ripke 1993) and by means of the LANDSAT multispectral images of 1972 and 1973 (Photo 6.1) the area was divided into the following large landscape units: Serirs, Ergs, Deflation plains, Deflated depression ("Deflationswannen"), hilly landscapes, plateaus with escarpments, wadis, oases, and birs. With the help of the geological map of southwest Egypt with a scale of 1:500,000 (Klitzsch and List 1980), the landscape units were divided into smaller geomorphological units. In the following chapters, each landscape unit and its geomorphology are presented in the order of decreasing areal size.

## 6.1 Serirs

Serirs are wide monotonous plains of alluvial sand and aeolian sand of different thicknesses (Photo 6.2). The surface of the Serir is usually covered with a layer of coarse sand to fine gravel. The Serirs are mainly very flat landscapes and some localities are weakly undulated. The monotony of the Serir is occasionally interrupted by dunes (usually barchanes) or by outcrops of quartzite sandstone or iron crust remnants that appear from far like small black islands in a sea of sand. The most famous Serirs in the studied region are: "Atmur El Kabeish" and the "Selima Sand Sheet". The limits between Atmur El Kabeish, the Selima Sand Sheet, and the other Serire are difficult to draw.

The Selima Sand Sheet is the largest serir area. It covers an area of 60,000 km$^2$ in southern Egypt and North Sudan. The Egyptian/Sudanese border divides this serir into two parts of almost equal size. In the Egyptian part, the terrain falls gradually to the east and north. In the east, we find numerous Birs. The most famous Birs are: Bir Sahara, Bir Misaha, Bir Tarfawi, and Bir Safsaf. In contrast to Bir Tarfawi and Bir Safsaf, the first two birs are without vegetation. The lowest spots of the "Selima Sand Sheet" are at Bir Tarfawi and Bir Safsaf. There, the terrain surface is 250–300 m

**Photo 6.1** LANDSAT I Composite satellite images of southern Egypt taken in 1972 and 1973. Top right is the southern tip of the Sinai Peninsula, in the center, the Abu Tartur escarpment, and at the bottom left is the plateau of Gilf Kebir

**Photo 6.2** The Selima Serir with air reflection (Mirage) at noon time

**Photo 6.3** Parts of the Serirs are covered with loose sand and cannot be crossed by car without aluminum sheets

above sea level and rises west to 600 m above sea level at the Gilf Kebir. From the west and northwest toward Bir Tarfawi and Bir Safsaf, the depth of the groundwater level decreases. The groundwater occurs at a depth of a few centimeters at Bir Tarfawi and at a few meters' depth at Bir Safsaf.

In general, Serirs are easy to drive. However, some locations are covered with loose aeolian sand and are not easy to go by car (Photo 6.3). Also, other locations are covered by dense and hard sand ribs. The thickness of the sand ribs ranges from a few centimeters to a few decimeters (Photo 6.4). On such ground surface, vehicles will shake vigorously and make driving uncomfortable. Rarely ground surfaces with parallel sand waves with approx. 2 m wavelength and 50 cm wave-amplitude (Photo 6.5) was found on a Serir landscape. On such sand waves, driving perpendicular to the waves can swing the cars in an unbearable way.

## 6.2 Ergs

The second largest geomorphological unit is the Ergs. They are characterized by aeolian sand and dunes deposited on different substrates (Photo 6.6). Almost all types of dunes can be found in the investigated region: barchans (creeping dune or crescent dune, Photo 6.7), Seifs (longitudinal dunes, Photo 6.8), dome dunes (Photo 6.9), and transition forms. According to the main wind direction Barchans have the lee side on

**Photo 6.4** In some localities, the surface of the Serir has hard sand ribs. When driving in such areas, the cars are heavily shaken

**Photo 6.5** Hard undulating ground surface that stretches for some kilometers and causes an unpleasant rocking of moving cars

**Photo 6.6** The Great Sand Sea in west and south-west Egypt is the largest Erg of the investigated region

their southern part and they move in the same direction and the longitudinal dunes (Seifs) are directed from North to south. Seifs are usually deposited on long aeolian sand waves. Such aeolian sand waves are called whale back. Barchanes characterize mainly the southern portions of the Ergs.

The largest Erg landscape is the Great Sand Sea. It covers an area of 70,000 $km^2$. It consists mainly of Seifs of about 100 km in length, up to 500 m in width, and partly over 70 m in height. Such dunes are easily recognized on satellite images. The thickness of the accumulated sand in the Great Sand Sea is 120–300 m (Haynes 1982). Occasionally, the hard rock of the underground as well as playas occurs between the rows of Seifs.

The second largest Erg is located east of the oases chain Kharga-Baris. It is much smaller than the Great Sand Sea and represents the lower third of Erg Abu Maharik. The road of Kharga-Dakhla as well as the road of Kharga-Baris are often closed by creeping dunes (Barchanes) of this erg and the traffic has to be often diverted (Photo 6.10). Measurements showed that barchans of 200 m length, 275 m width and 20 m height (approx. 760,000 t) move at an average speed of 7.5 m/year (Haynes 1989).

On the Serirs and Deflation areas, smaller ergs of 1–2 km wide and a few kilometers length can also be found. They mainly contain barchans. Dome dunes are rare and have only been observed on a serir.

**Photo 6.7** Barchans (sickle dunes, migratory dunes, creeping dunes, crescent dunes) occur mainly in the south from Ergs

**Photo 6.8** Seifs or longitudinal dunes are common in Ergs. The picture shows a Seif of more than 100 km in length (at the base of the dune is an Off-road vehicle and an Unimog presented as two small points)

**Photo 6.9** Dome dunes occur in areas with predominantly high wind speeds

**Photo 6.10** The road Kharga–Dakhla is often blocked by Barchans and has to be frequently diverted

## 6.3 Deflation Areas

Deflation areas are landscapes formed by wind erosion. They can be divided into Deflated plains and Deflated depressions (Deflationswannen). Deflated plains occupy large areas in southwestern Egypt. According to the main wind direction, the deflated depressions are mainly found south of the Egyptian and Abu Ballas escarpment. Its formation is largely due to the erosional effect of fall winds (which blows from the plateaus and fall down the escarpment). This is supported by Yardangs, which occur in the deflated depressions west of Dakhla and in the foreland of the Abu Ballas escarpment. In both locations, the Yardangs are lined up toward the escarpment, from where the wind blows. During our expeditions in these locations (near Abu Ballas and Dakhla) we could observe such strong fall wind and wind gusts blowing from the escarpments every night. The force of the wind could be felt and heard through the shaking of our vehicles and the impact of sand on metallic objects.

Due to the selective erosion of wind, the ground surface of deflated areas is covered with a layer of stones and/or fine gravel. The stones and occasionally the fine gravel can consist of iron crusts, iron concretions (iron oolite), or silcrite residues. The ground surfaces of some deflation plains are covered with a dense cover of middle gravel to coarse stones. Such deflation plains are known as hamada. Also, Serir landscapes could be included to deflation plains. But in this book, deflation plains are only those landscapes that consist of tertiary, Mesozoic, or crystalline rocks and do not occur on plateaus. Those developed on solid rocks have a rougher micro-relief compared to Serir (Photo 6.11). They have small dents, thresholds, steps, and large blocks of stones that are residual due to their great hardness and resistance to wind erosion (Photo 6.12). Deflation plains occur on almost all tertiary and Mesozoic rocks. The largest deflation plains are on Six Hills, Sabaya, and Duwi Formations, as well as on crystalline rocks north of Bir Safsaf.

The largest deflated depression, which has no groundwater connection, is located in the foreland of Abu Ballas escarpment (Photo 6.13) and is formed out of the formation of the same name. Its deepest part is covered by several playas, which have mighty Yardangs and are interrupted by several dunes. The Yardangs are formed like giant shark fins up to 5 m height (Photo 6.14). On a neighboring playa, they are shaped like giant frogs (Photo 6.15). In the foreland of Abu Tartur escarpment and between Dakhla and Baris, the Yardangs consist of elongated stripes (Photo 6.16).

Deflated depressions with groundwater contact are classified separately as oasis landscapes.

## 6.4 Hilly Landscapes

Most hilly sites in southwest Egypt are remnants of plateaus and escarpments. The hills are "inselbergs" and "tafelbergs" from tertiary (in northeast) and Mesozoic rocks (Photo 6.17). Its height is a few meters to a few hundred meters. In addition,

**Photo 6.11** Deflation plain on hard Mesozoic rock

**Photo 6.12** Deflation plain on white sandstone with residual rock blocks

**Photo 6.13** Playa sediment with Yardangs in a deflated depression in foreland of Abu Ballas escarpment

**Photo 6.14** Yardangs out of playa sediment in the form of giant shark fins in foreland of Abu Ballas escarpment

**Photo 6.15** Yardangs in frog form in foreland of Abu Ballas escarpment

**Photo 6.16** Longitudinal Yardangs in the foreshore of the Egyptian northern plateau

**Photo 6.17** Hilly landscape with inselbergs and tafelbergs from Mesozoic rocks

basalt mounds (cones) and rarely quartzite sandstone mounds occupy small areas. The domes and slopes of the hills are covered with sharp-edged stones and rocks blackened by desert varnish as well as iron crusts. Due to the desert varnish on the stone surfaces and/or iron oxides, almost all hills have a black color (Photo 6.18). Between the hills, playas, fluvial sediments, and aeolian sand are often found in different thicknesses. Small dunes or aeolian sand sheets can occur on the lee side of the hills. The latter can be completely covered by debris. The most famous hills are Gebel Edmonstone (468 m a.s.l.), Six Hills (approx. 450 m a.s.l.), Abu Ballas (467 m a.s.l.) and Gebel Kamil (785 m a.s.l.).

*Gebel Edmonstone* is located west of Dakhla. It rises for about 200 m from its surroundings, represents a tafelberg, and consists of the Tarawan Formation. The *Six Hills* consist of six conical iselberge. They are about 100 m high and consist of the Six Hills formation, which represents the Cretaceous/Jura transition. The greatest number of hills occurs in the area of the Abu Ballas Formation. The area of Abu Ballas (in Arabic means the father of pots) is the most famous inselberg in the investigated area. It is called the *Abu Ballas* Inselberg. At the foot of this inselberg there are shards of several hundred pots, which are dated for approx. 1200 years BC. The Abu Ballas Inselberg is about 50 m high and consists of the rock-sediments with the same name. *Gebel Kamil* is also an inselberg; it is about 100 m high and consists of rocks of the Gilf Kebir Formation. It is located on the border with Sudan, southeast of the Gilf Kebir. During the last World War, Gebel Kamil was in the flight path of a former Anglo-Canadian military airport.

**Photo 6.18** Hills blackened by desert varnish

## 6.5 Plateaus with Escarpments

In the investigated region there are three plateaus: the Egyptian Plateau, the Abu Ballas, and the Gilf Kebir Plateau. The slopes of escarpments found at the edge of plateaus are covered with debris which leis over pediments of different widths. The edges of the plateaus and escarpments as well as the pediment areas are rugged and fissured by wadis and dry water channels.

The *Egyptian plateau* is located in the north and east of the investigated region and extends northward to the Mediterranean Sea. Within the investigated area, this plateau can be divided into two large parts; the northern part (north Dakhla-Kharga) and the eastern part (east Kharga-Baris). The northern plateau also includes the Abu Tartur Plateau. In Erg areas (e.g., north of Kharga near Abu Tartur and west of Dakhla) the slopes are covered by aeolian sand. Otherwise, all slope foot has thick debris cones. The altitude of the North Plateau decreases toward the west. It is between 448 and 490 m above sea level (Photo 6.19). The altitude of the eastern plateau decreases toward the south and is between 420 and 275 m above sea level. The escarpment (Cuesta) of the northern plateau is about 300–400 m high. It consists of siltstone, sandstone, and limestone of the Tarawan, Kurkur, and Dakhla Formation. The eastern plateau is much less high. Its escarpment is divided into several small stages, all of which are made of limestone. The eaves and plateau of the north and east escarpment are mostly made of limestone. Moreover, Karst phenomena can be observed on the plateau (Kröpelin 1993).

**Photo 6.19** On the edge of Abu Tartur Plateau (approx. 400 m high): Wadis can be seen in the foreland

The *Gilf Kebir Plateau* is much smaller than the Egyptian plateau. It was first discovered in 1920 or 1923 by the Egyptian King Fuad's chamberlain Hassanein-Bey (Hans Rhotert 1978). It is 12,000 km$^2$ in size, almost oval, and consists of sandstones and siltstones of the Sabaya Formation. Some basalt cones are found scattered on the plateau. The height of the plateau is about 1050 m above sea level in the south, and about 900 m above sea level in the north. Its altitude is on average 300 m and decreases to the north (Photo 6.20). The escarpment of the Gilf Kebir Plateau is very steep, especially in the south and in the east. The lower part of the slopes is often covered with aeolian sand. The upper part of the slopes is almost vertical and consists of hard sandstones. The Periphery of the Gilf Kebir plateau is cut by several deep wadis. In the northeast, the plateau diminishes into numerous tafelbergs and inselbergs.

Northeast of the Gilf Kebir Plateau is the *Abu Ballas Plateau*. Its escarpment is 50 km long, 100 m high, and facing west/east. The slopes of the escarpment are very steep, and often almost vertical (Photo 6.21). The eaves and the surface of the plateau consist of hard sandstones of the Sabaya Formation. Similar to the Gilf Kebir, the lower slopes are often covered with aeolian sand. Compared to the Egyptian plateau and the Gilf Kebir, the Abu Ballas escarpment shows nearly no fluvial morphology (like wadis or dry water channels). It seems that wind erosion has blurred every sein of running water activities. The roof area is characterized by many small hills. To the north, the hills are becoming less and the landscape is slowly changing into a plain covered by aeolian sand.

**Photo 6.20** Escarpment of the southern part of the Gilf Kebir Plateau near Wadi Wassa

**Photo 6.21** Escarpment of Abu Ballas Plateau with playa sediment in the foreland

**Photo 6.22** Hamada landscape on the plateau of the Gilf Kebir with circular-shaped hollows. The stones on the ground surface are blackened by desert varnish

All plateaus are covered with stones of different sizes, which are blackened with different intensity by desert varnish. The plateaus are typical hamada landscapes (Photo 6.22). However, the black color of desert varnish on limestones of the Egyptian plateau is much weaker than the black color on stones of silicates found on the Gilf Kebir and the Abu Ballas plateau. In general, the plateaus are plains with a rough micro-relief. Small hollows on the Gilf Kebir and the Egyptian plateau are often filled with colluvium.

## 6.6 Wadis

The wadis (ephemeral valleys) in the studied region were developed during Tertiary and Pleistocene (Photo 6.23). They are found on the periphery of the Egyptian and the Gilf Kebir plateaus and extend to the foreland of their escarpments. In the local language they are called Khiran (singular: Khor).

Under present climate conditions, water flow in the wadis is very rare. Depending on episodic rains water flow in the wadis is approximately once in every decade. The amount of water from such episodic precipitation is often not sufficient to supply the entire water channel with water. Usually the surface water seeps after a short flow distance. From the recent dry water channels, it can be concluded that the water seeps after a few hundred meters. Because of the uneven water flow, the wadis show

**Photo 6.23** Wadi Dayek in the eastern part of the Gilf Kebir Plateau

only partially balanced inclination. Wadis, which have had water in recent decades, occasionally show some plants that were mostly dead during our visits. Compared to the Gilf Kebir and the Egyptian escarpment, the Abu Ballas escarpment shows no recent influence of running water.

In the Gilf Kebir, the wadis, especially in the south and east, have cut deep gullies in the plateau. They are bordered by very steep slopes with an average slope angle of 35° which become steeper to the top and culminate in a vertical or nearly vertical angle (Photo 6.24). In the Gilf Kebir, wadi terraces are usually not present. An exception is Wadi Maftuh, which has a terrace 100 m wide and 3 km long (Photo 6.25) (Kröpelin 1989). In the area of the Egyptian plateau, three to four terraces were observed in a wadi east of Bulaq (Smettan 1987).

In the upper reaches of wadis, there is usually loose sand, in the lower reaches some densely accumulated fine-layered silty sand and playa sediments can be found in final pans. The latter are located in the foreland of escarpments or often in deflated depressions. Several west/east-facing wadis are blocked by barrier dunes. A barrier dune of several meters in height is located in Wadi Maftuh (Photo 4.3). Behind such barrier dunes, silty water sediments a few centimeters thick and playa sediments of several meters thick can be found, such as, e.g., in Wadi Bakht and Wadi Akhdar (Photo 6.26).

The most famous wadis in the area of Gilf Kebir are Wadi Wassa, Wadi Akhdar (green), Wadi El Gazaer (islands), Wadi Bakht (lucky), Wadi Maftuh (open), Wadi Dayak (narrow), Wadi Contrast, Wadi Abd El Malik (kings-slave. This can be the name of a person). (In brackets the meaning of the Arabic names.)

**Photo 6.24** Some Wadis in the Gilf Kebir have steep cliff walls

**Photo 6.25** Wadi Terrace in the lower reaches of Wadi Maftuh in the east of Gilf Kebir

**Photo 6.26** Playa sediment behind a barrier dune in Wadi Akhdar (south of Gilf Kebir)

With the exception of Wadi Battikh (melon), Wadi Naqb El Rufuf, and Wadi El Abd (slave), most of the wadis of the Egyptian escarpment are unnamed on topographical maps. Wadi Battikh is located west of the Abu Tartur Plateau, Wadi Naqb El Rufuf northeast of Kharga, and Wadi El Abd east of Baris.

## 6.7 Oases

The inhabited oases lie in two major depressions: the Kharga-Baris and The Dakhla Depression (Photo 6.27). In these depressions, there are a large number of wells and settlements. The Kharga-Baris Depression is located 150 km west of the Nile Valley. It is laid out in a north–south direction, which is due to a tectonic rift (Schiffers 1973). The depression covers 3693 $km^2$ and according to Kröpelin (1993), its largest length is 185 km, while its width varies between 20 and 80 km. It has an average altitude of 38–80 m. a. s. l. and it reaches sea level at its lowest points (Smettan 1987). The depression is bordered in the east and northwest by the escarpment of the Egyptian plateau, which is built from mighty tertiary sediments. The Egyptian Plateau northern from Kharga rises up to 400 m above the oasis floor. To the west is Abu El Mahariq Erg. The heights of the barchans of this erg are on average 8–12 m. These shifting dunes move with an estimated speed of 12 m $a^{-1}$. As a result, the Kharga/Asiut, Kharga/Baris, and Kharga/Dakhla routes are often blocked by Barchans (Photo 6.10). The Kharga-Baris Depression is largely found within the

**Photo 6.27** The oases are found in deflated depressions. The picture shows one of the oases of Kharga

Mut Formation. In the north and west of Kharga occur the Maghrabi and the Taref Formation. The Mesozoic rocks are mostly covered with aeolian sand in different thicknesses. Playa sediments occur especially in Baris, Bulaq as well as in smaller areas south and north of Kharga. North of the oasis Baris, long Yardangs have been modeled out of playa sediments by wind erosion (Photo 6.28).

The Dakhla Depression is located 120 km west of the Kharga Oases. It is significantly smaller than the Kharga-Baris depression. It extends in WNW to ESE direction, with a length of 50 km, a width of 20 km, and an area of about 2000 km$^2$. The depression of Dakhla is between 90 and 140 m above sea level and is crossed by Barchans from a small north/south stretched erg (Photo 6.29). In the lowest part of the depression, playas are found. Yardangs from playa sediments occur west of El Qasr. The underground of the Dakhla Depression consists of the Maghrabi and Taref Formation. The geological map of SW Egypt shows Dakhla Shale and Plant Bed for the oasis area. The playas and the rocks of the Maghrabi and Taref Formations are partially or completely covered with different amounts of aeolian sand.

In the winter season of 1981/82, the arable area in Kharga was 3893 ha and in Dakhla 9022 ha. At Kharga there were 1050 deep-water wells and 504 shallow wells, at Dakhla 2520 deep wells and 2226 shallow wells (Beller Consult 1985). According to the Geopotential Maps of the Western Desert of Egypt (Meissner et al. 1993):

- The Baris Oasis consists of one Village (Baris) and five Ezbas (=Farmsteads),
- The Kharga Oasis consists of one Town (El Kharga), three Villages and eight Ezbas, and

**Photo 6.28** According to the wind direction, the longitudinal Yardangs are directed North/South at the oasis Baris

**Photo 6.29** Barchans of an Erg near Dakhla in foreland of the Egyptian Plateau

- The Dakhla Oasis consists of 14 Villages and 13 Ezbas.

In addition to the arable areas and settlements, there are so-called evaporation ponds or drainage ponds surrounded by sebchas in the lowest part of the depressions in the vicinity of farmland.

## 6.8 Birs

In Arabic, Bir means waterhole = water well. Most of the Birs are locations with high groundwater levels, and have wild vegetation and wild animals but are not inhabited by humans (Photo 6.30). Birs are found in south-east of the investigated region. They are located between 258 and 176 m above sea level. The most famous Birs are: Bir Tarfawi, Bir Safsaf, Bir Sahara, and Bir Misaha. All four Birs occur in the lowest part of the Selima Sand Sheet. Bir Sahara and Bir Misaha consist of deep-water holes and are vegetation-free. On the other hand, Bir Tarfawi is rich in vegetation especially tamarisks and date palms. It is 9 km long, about 1.5 km wide, and about 1350 ha in size (Schneider 1990). Bir Safsaf is 3.5 $km^2$ large and is covered with reeds as well as a bush of Dom palms and date palms (Darius 1989). With the exception of Bir Sahara and Bir Misaha, the groundwater at the other Birs lays a few decimeters underground and the soils are heavily enriched with salt and/or gypsum. At Bir Sahar and Bir Misaha, groundwater can be found at a depth of approx. 30 m and 18 m, respectively (Thorweihe and Heinl 1990).

**Photo 6.30** Groups of dom palms at Bir Mur

In southeast, outside the studied region are Bir Nakheila, Bir Mur, Bir Abu El Husein (without vegetation), Bir Kiseiba, Bir Kurayim, Bir El Shab, Bir Takhlis, and Bir Nakhlai. All this Birs are located on Mesozoic rocks.

## References

Beller Consult Ltd (1985) Water resources and soil potential development project New Valley. Freiburg/Siegen

Darius F (1989) Wasser- und Nährstoffhaushalt einer unbewohnten Kleinoase Süd-Ägyptens. Diss. TU-Berlin, p 83

Hans Rhotert S (1978) Die Erforschung der Wüste. In: Museen der Stadt Köln: Sahara – 10000 Jahre zwischen Weide und Wüste, pp 22–34

Haynes F (1982) Great Sand Sea and Selima Sand Sheet, Eastern Sahara: geochronology of desertification. Science 217:627–633

Haynes F (1989) Bagnold's Barchan: a 57-yr record of dune movement in the Eastern Sahara and implications for dune origin and paleoclimate since Neolithic times. Quatern Res 32:153–167

Klitzsch E, List FK (1980) Egypt-geological interpretation map, 1:500,000 (preliminary edn). TFH Berlin

Kröpelin S (1989) Untersuchungen zum Sedimentationsmillieu von playas im Gilf Kebir (südwest-Ägypten). Africa Praehistorica 2:183–305

Kröpelin S (1993) Geomorphology, landscape evolution and Paleoclimates of Southwest Egypt. Catena Supplement 26:31–65

Meissner B, List K, Ripke U (1993) Topography map, South-West Egypt 1: 1000000. Catena Supplement 26

Meissner B, U Ripke (1993) The use of remote sensing data in cartography, examples from Egypt and Sudan. Catena Supplement 26:5–29

Schiffers H (1973) Die Sahara und ihre Randgebiete III. Weltforum, München

Schneider U (1990) Pflanzenökologische Untersuchungen an Wild- und Kultuepflanzen in der Extremwüste Südägyptens. Diss. TU-Berlin, p 292

Smettan U (1987) Typische Böden und Bodengesellschaften der Extremwüste Südwest-Ägyptens. Berliner geow. Abh. A 83:1–190

Thorweihe U, Heinl M (1990) South-West Egypt hydrolgeology map 1:1,000,000. In: Meisner B, Wycisk W (eds) Geopotential and ecology—analysis of a desert region. Catena Supplement 26, Cremlingen.

# Chapter 7
# General Soil Properties

**Abstract** Under the high and long-lasting aridity, all terrestrial soils are extremely dry. The water content of terrestrial soils is well below the unavailable water capacity. Due to the lack of vegetation, the soils are almost free of organic material and have no visible humus horizon. Most soils have weak reddish to slightly brownish B-horizon with different thicknesses. The soils are usually covered with a layer of aeolian sand and/or stones. The top soil of hamadas is completely covered with medium gravel to coarse stones, while on serirs they are covered with coarse sand to fine gravel. The stones and gravel are covered with desert varnish. Some stones on the soil surface are polished by wind and have a shiny surface, which is also called desert varnish or patina. Under the stones, we often find a patch of white powdered gypsum. The stones and the sand cover lie on a thin silty layer of a few mm thicknesses. This layer shows fine cracks and contains different amounts of carbonate, while the sand cover and the upper soil horizon can be free from carbonate. Under the silty layer occur crack polygons filled with sand. In the center of the polygons, there is often a patch of gypsum. Some mega-cracks reach depths of more than three meters and have widths at the soil surface from a few centimeters to a few decimeters. In addition, almost all upper soils have a platy structure. In the case of loamy upper soils, the platy structure has often a spongy or foamy structure. The subsoils have a very high bulk density. The genesis of most field properties is described in this chapter and some phenomena that are difficult to explain are presented in the chapter. Outside oases, birs, and wadis, the average organic material content is 0.18% and the nitrogen content is 134 ppm. The nitrogen is present in inorganic form. Due to the high solubility of nitrates, the highest amount is found in the subsoil from terrestrial soils. Groundwater soils are almost free of inorganic nitrogen. The chemical analysis showed that terrestrial soils contain different amounts of soluble salts, gypsum, and carbonates. In general, the content of soluble salts increases with increasing clay content. In terrestrial locations, the soluble salts and gypsum are usually enriched in the subsoil, the carbonate in the topsoil. The pH of salt-rich samples is usually between 7.0 and 8.5, while the pH values of samples with low-salt content and salt-free samples vary between 6.5 and 10.0. The clay fraction of soils from Mesozoic sediments and quaternary sediments in southwest Egypt consists mainly of kaolinite.



F. Alaily, *Ecosystems of Extremes*, Advances in Geographical and Environmental Sciences, https://doi.org/10.1007/978-3-031-76910-8_7

**Keywords** Hamada · Serir · Gypsum · Desert varnish · Cracks · Sand cover · Platy structure · Nitrates · Soluble salts · pH values · Spongy structure · Foamy structure

This chapter presents terrestrial soil properties that are widespread and independent of soil type or soil parent material. These properties are mainly controlled by the climate conditions of the region.

## 7.1 Field Properties

Under the high and long-lasting aridity of central Eastern Sahara, all terrestrial soils (up to 2.00 m depth) were extremely dry during our visits in the studied region. Figure 7.1 shows that the water content of the studied terrestrial soils is well below the unavailable water capacity estimated according to Schlichting and Blume (1966) (pF > 4.2). The moisture of these soils has been described in the field as very dry and it must be bound with over 5000 cmWC (pF 5). With such binding force, soil moisture is not available even for desert plants. Due to the lack of vegetation, the soils are almost free of organic material and have no visible humus horizon. In addition, it can be observed that most soils have weak reddish to slightly brownish B-horizon with different thicknesses. Such a horizon can only arise if the soil is occasionally moistened and the crystalline iron of minerals is oxidized in amorphous type (Alaily 1986).

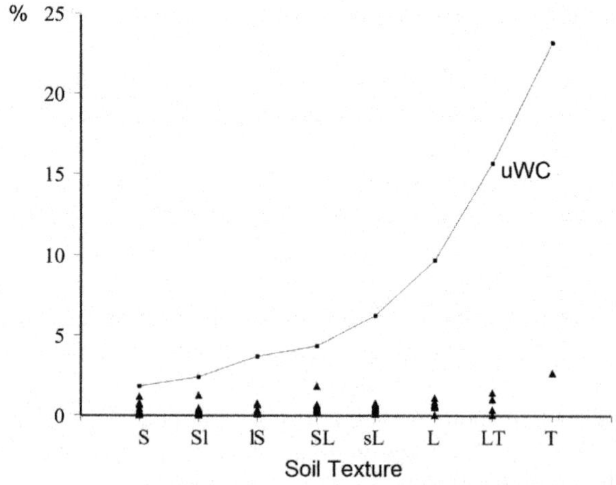


**Fig. 7.1** Water content and soil texture of terrestrial soil compared to unavailable water capacity (pF > 4.2)

**Photo 7.1** The soils in the studied region are usually covered with aeolian sand and/or stones, which are almost always covered by black or dark brown desert varnish

With the exception of some clay-rich playas, the soils are usually covered with a layer of aeolian sand and/or stones (Photo 7.1). The top soil of hamadas is completely covered with medium gravel to coarse stones, while on serires they are covered with coarse sand to fine gravel. The stones and gravel are covered in different intensities with desert varnish (also called desert polish or desert patina).The desert varnish is mostly black, occasionally brown. It consists of a thin skin of iron and manganese oxides. Some stones on the soil surface are polished by wind and have a shiny surface (Photo 7.2), which is also called desert varnish or patina. Under the stones, we often find a patch of white powdered gypsum (Photo 7.3).

The stones and the sand cover lie on a thin silty layer of a few mm thicknesses (Photo 7.4). This layer shows fine cracks and contains different amounts of carbonate, while the sand cover and the upper soil horizon can be free from carbonate. The carbonate content of two silty layers analyzed in the laboratory was 2.5 and 7.7%. The grain size analysis shows that compared to the sand cover and the top soil, the silty layer has a much higher clay and silt content (Fig. 7.2). The silty layer under the sand cover can be explained by the high cohesion force of silt particles and repeated deposition of aeolian material on the soil surface, followed by selective erosion of sand without silt. Due to this process, as well as the absence of plants and soil fauna,

**Photo 7.2** The stones on the soil surface are often shiny and polished by wind-sand-blowers

**Photo 7.3** One often finds a gypsum stain under the stones of the soil cover. The photo shows an upturned stone with white gypsum patches

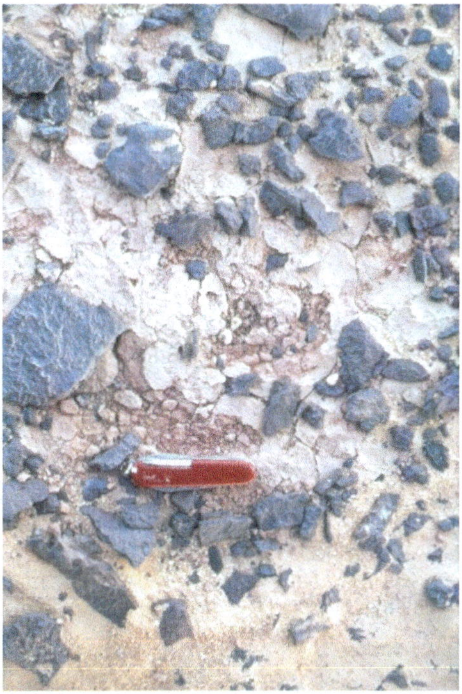

**Photo 7.4** Under the aeolian sand cover there is a 1 mm thin silty layer

the silty layer could accumulate without disturbance in the course of time. This means that the sand of the soil cover is replaced at each sandstorm, while silt and clay are stuck on the soil surface and will be added in small portions. This consideration derives from the fact that the wind can blow away sand more easily than silt or clay, which adheres more strongly to the soil surface by cohesion forces. Weak episodic rain and salt cause the particles to combine and form a relatively coherent layer.

Under the silty layer occur crack polygons filled with sand. In the center of the polygons, there is often a patch of gypsum with a few millimeters to one-centimeter thickness (Photos 7.5 and 7.6). The cracks, which occur in almost all soils (Photos 7.7 and 7.8), even in sandy soils, appear in large numbers as fine cracks (marked with A in Fig. 7.3) in the top 20 cm. Up to a depth of about 40 cm, numerous and larger cracks (shown with B in the figure) of a few millimeters wide occur. Some mega-cracks (shown with C) reach depths of more than three meters and have widths at the soil surface from a few centimeters to a few decimeters.

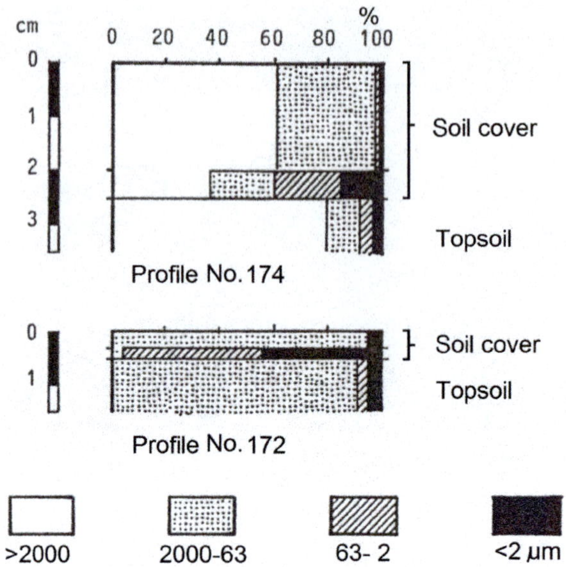


**Fig. 7.2** Grain size distribution of the silty layer under the aeolian sand cover

**Photo 7.5** On the soil surface gypsum is enriched in the center of small polygon cracks

**Photo 7.6** The gypsum spot in the center of the polygon cracks is only a few millimeters thick

**Photo 7.7** A representative soil profile of Central East Sahara with sand-filled cracks which are decimeter wide and several meters deep

**Photo 7.8** Sometimes large crack polygons (diameter approx. 1.5 m) can be seen on the ground surface

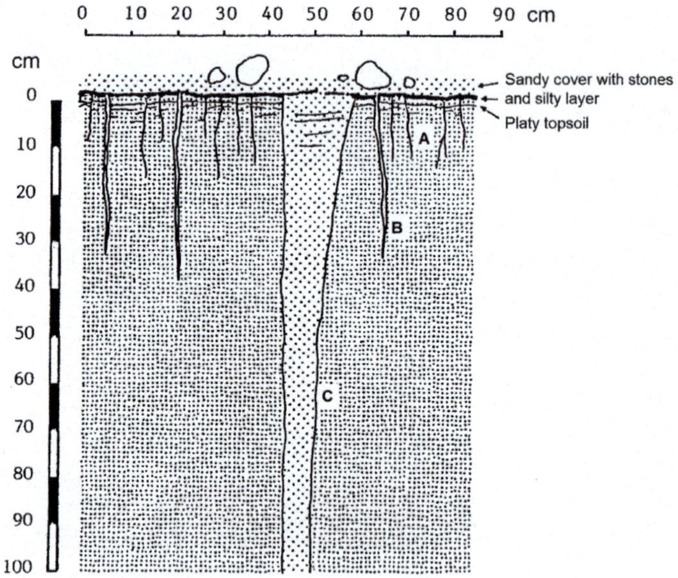


**Fig. 7.3** Ideal soil profile of the extreme desert with typical and frequent macro-morphological properties (A = fine cracks, B = small cracks, C = mega-crack)

**Photo 7.9** Some cracks are hollow and sealed with soil aggregates

The mega-cracks (C) are even found in sandy soils and are often filled with fine stratified sand. Occasionally, mega-cracks were observed sealed by soil aggregates and were partially empty (Photo 7.9). Due to vibration (e.g., by moving vehicles) the surface sand can trickle into the empty cracks, creating small Doline-shaped depressions on the soil surface (Photo 7.10).

In addition, almost all upper soils have a platy structure. In the case of loamy upper soils, the platy structure has often a spongy structure or a foamy structure, which is characterized by air bubbles (Photo 7.11). Soils covered with stones usually have a more pronounced foamy structure than those covered only with aeolian sand. In addition, upper soils with a pronounced foamy structure are mostly calcareous, so the genesis of the foamy structure could be related to the enrichment of carbonate. Table 7.1 shows that only 4.9% of the foamy structure was free of carbonate. Otherwise, most samples had Carbonate content between 2 and 11%.

The platy topsoil structure as well as the cracks is missing in very young soils of recent sediments (e.g., aeolian sand and recent fluvial sediments). In addition, the subsoils have a very high bulk density, so it is not possible to drive the Puerckhauer (Hollow metal rod used for soil examination) into the soil. Blume and Petermann (1979) made similar observations for sandy soils of the Fezzan in Libya. The high bulk density of the subsoil is not only due to a lack of biological activity. It is due to

**Photo 7.10** Due to ground vibration, sand from the soil cover pours into the hollow cracks, and small dolines are formed on the soil surface

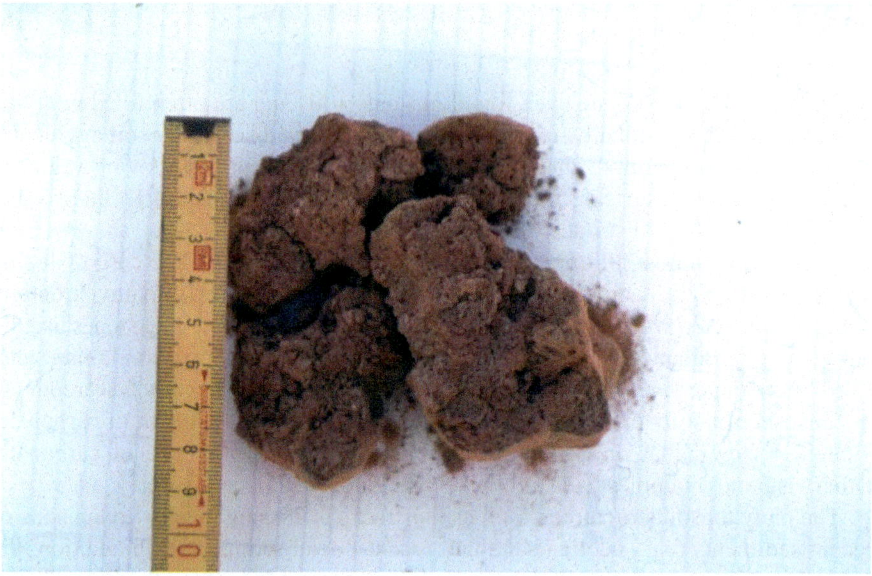

**Photo 7.11** The platy foamy topsoil structure is relatively widespread in loamy soils of Central East Sahara

**Table 7.1** Carbonate content of the foamy soil structure estimated in the field with 10% HCl

| Estimated carbonate content | Carbonate | | | | | |
|---|---|---|---|---|---|---|
| | Free | Very poor | Poor | Containing | Rich | Very rich |
| Number of samples | 11 | 8 | 31 | 94 | 69 | 10 |
| Number of samples in % | 4.9 | 3.6 | 13.9 | 42.1 | 30.9 | 4.5 |

the strong drying of clay-rich soils and in sandy soils, it is caused by the arrangement of dry sand toward maximum bulk density as a result of expanding and shrinkage due to daily and annual temperature fluctuations (see cracks genesis).

The above-mentioned properties are widely spread in the studied region. They cannot all be considered as typical properties of soils of the extreme desert, as some of these properties occur in other soil zones. Typical pedogenous properties of the extreme desert are only:

(1) Desert varnish, (2) gypsum patches under the stones of the soil cover, (3) silty topsoil cover, (4) foamy topsoil structure, and (5) the extremely high dryness of the subsoils.

In addition to the above-mentioned phenomena, some observations have been made which are difficult to explain. These are described below.

It was often observed in locations with very shallow soil (Leptosol sites) and some platy stones upright on the soil surface (Photo 7.12a). Photo 7.12b shows vertically standing stone slabs arranged in straight lines, while Photo 7.12c shows vertically standing rock thin plates arranged in circles. Similar phenomena have been described in Arctic regions on permafrost soil as well as in the savannas on Vertisols. In the first case, they are caused by cryoturbation, in the second case by peloturbation. Since the soils are poor in clay and periglacial climatic conditions did not occur during the Quaternary, the mechanisms for the formation of such phenomena remain unclear for the subtropical extreme desert.

In addition, two rare observations were made, which are documented in Photos 7.13 and 7.14. The cause of the parallel sliding tracks (Photo 7.14) on a slope is not found at the lower part of the slope. It is unlikely that such traces were caused by the slide of rock blocks, as stones bounce and roll on such steep slopes as they move downhill. Therefore, their tracks would not be continuous and would not form a straight line. In addition, no corresponding stones were found on the slope foot. The second observation in Photo 7.14 is also difficult to explain. Due to the pressure of a moving vehicle, the aeolian sand, which was relatively freshly deposited in an older tire lane, was lifted up Board-like. The board-like lifting of the sand could indicate that the loose sand was compacted and hardened by salt and/or cohesion forces in a relatively short time.

**Photo 7.12** Vertical rock plates: **a** standing individually, **b** in parallel lines, **c** arranged in ring form

**Photo 7.13** Parallel slide traces on a hillside

**Photo 7.14** Board-like lifting of loosely deposited aeolian sand in an old car tire track

## 7.2 Chemical Properties

### 7.2.1 Nitrogen and Carbon Content

The soils of the studied region show very low nitrogen content and very different levels of carbon content. Outside oases, birs, and wadis, the average organic material content is 0.18% and the nitrogen content is 134 ppm. In the wadis, although there is some vegetation, the average amount of organic matter is slightly lower (0.12% $C_{org}$ and 89 ppm N), while in the oases and birs the values are significantly higher (0.89% $C_{org}$ and 269 ppm N).

Nitrogen and carbon are present in organic and inorganic substances. The inorganic carbon occurs as carbonate and is found in almost all studied soils. The relationship between total carbon and carbonate content in Fig. 7.4 shows that most of the carbon is found in carbonate. By subtracting the carbonate carbon (($CaCO_3$ × 12)/100) from the total carbon content ($C_t$), one gets the content of organic carbon ($C_{org}$):

$$C_t - ((CaCO_3 \times 12)/100) = C_{org}$$

The $C_{org}$ contents determined by the formula written above are shown in Fig. 7.5. The five samples with a high content of $C_{org}$ have been determined in soil samples from tamarisk hills. The lack of a relationship between $C_{org}$ and nitrogen content in Fig. 7.5 indicates that most of the nitrogen is present in inorganic form. In the terrestrial soils examined, nitrogen must be present as nitrate. This is supported by the positive relationship between electrical conductivity and nitrogen content (Fig. 7.6). According to Smettan (1987), the average nitrate nitrogen content in the fine earth is 91 ppm and the highest content is 1163 ppm. Due to the high solubility of nitrates, the highest amount is found in the subsoil from terrestrial soils. Groundwater soils are almost free of inorganic nitrogen.

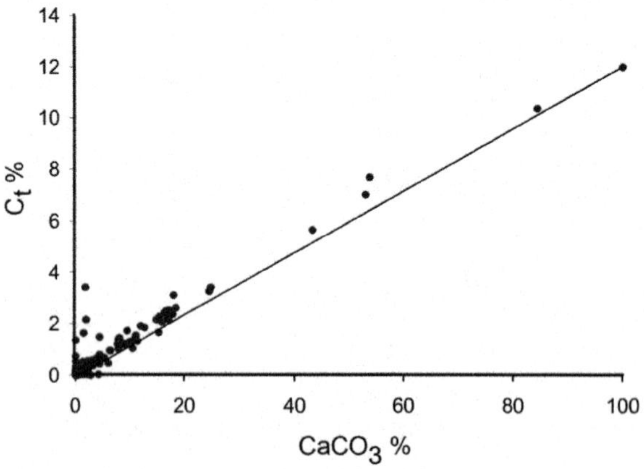


**Fig. 7.4** Relationship between total carbon and carbonate content in fine earth

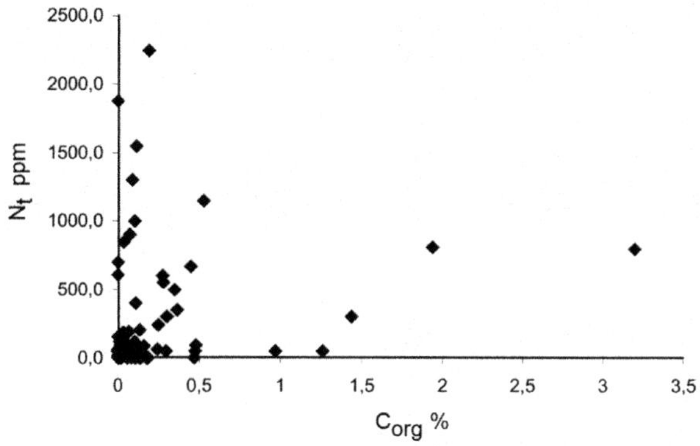


**Fig. 7.5** Relationship between organic carbon and total nitrogen content in fine earth

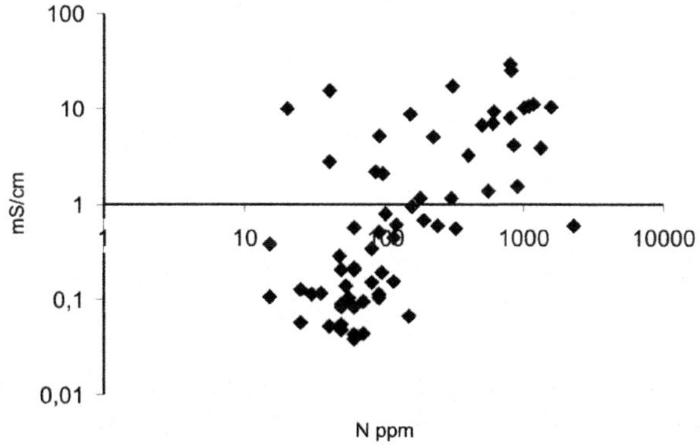


**Fig. 7.6** Relationship between log EC (electrical conductivity) and nitrogen content in the investigated terrestrial soil samples

## *7.2.2 pH and Salinity*

The chemical analysis showed that terrestrial soils contain different amounts of soluble salts, gypsum, and carbonates. The occurrence of gypsum and soluble salts under the stones on the soil surface can be explained by weak precipitation events or dew water (condense water). These will moist a few millimeters of the soil surface. When the water evaporates, the water around the stone is first lost, the evaporation front creeps under the stone and thus the dissolved substances (soluble salts and gypsum) move toward the stone center, where they precipitate. In addition, the stone protects the enriched salt and gypsum from being blown out by wind or from leaching by weak rainfall.

In general, the content of soluble salts increases with increasing clay content (Fig. 7.7). In terrestrial locations, the soluble salts and gypsum are usually enriched in the subsoil, the carbonate in the topsoil. In terrestrial soils, all three substances are atmogen (from dust) and/or lithogen, their depth function must have been caused by episodic rain (Alaily 1985; Alaily and Blume 1983; Blume et al. 1984; Blume

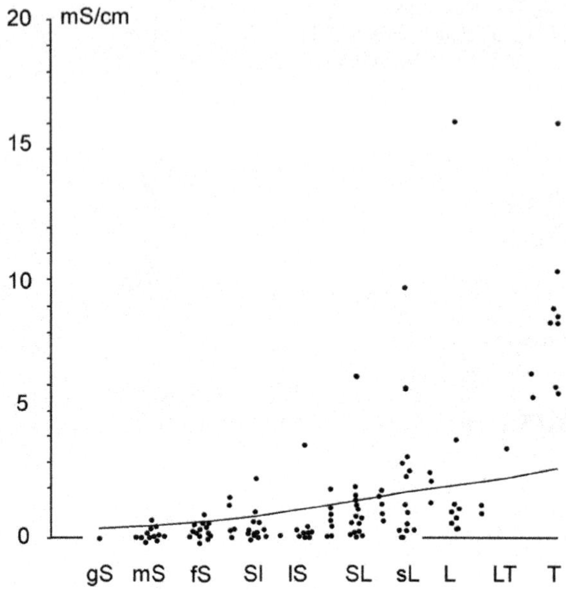


**Fig. 7.7** Relationship between soil texture and electrical conductivity of 1:2.5 soil:water extract from the investigated soils in southwest Egypt. The solid curve shows the 15 mS/cm equivalent EC value of the saturated extract

1986). Irrigation tests at four sites in the investigated area (near Tarfawi and Kharga) have shown that under the present climatic conditions episodic heavy rains with a maximum of 8 mm can penetrate the soil to a depth of maximum 15 cm. Therefore, salts that are enriched deeper in the soil must be relict (Blume 1985).

The pH of salt-rich samples is usually between 7.0 and 8.5 (Fig. 7.8), while the pH values of samples with low-salt content and salt-free samples vary between 6.5 and 10.0. The high pH values of the low-salt content samples could be attributed to high Na saturation on the exchange complexes. pH values lower than 5 (up to 1.8) were measured in the laboratory in some soil samples from Gr and Go horizons of a Gleysol from Bir Nakhla. Measurements of the same soil samples in the field showed pH values of 7. The reason for this discrepancy is the oxidation of sulphide when drying the samples in the laboratory and formation of sulfuric acid:

$$4FeS_2 + 15O_2 + 2H_2O \rightarrow 2Fe_2(SO_4)_3 + 2H_2SO_4$$
$$2Fe_2(SO_4)_3 + 8H_2O \rightarrow 4FeOOH + 6H_2SO_4$$

Figure 7.9 shows a positive relationship between water-soluble sodium and electrical conductivity (in 1:2.5 soil:water extract = $EC_d$). This indicates that the salts in the soil studied consist largely of sodium-bearing salts. By looking at $EC_d$, $EC_5$ (electrical conductivity of 1:5 soil:water extract), and the pH of the aqueous soil extract, the salt type can be roughly estimated. Due to the low gypsum content in the saturated solution (2.4 g/l, Scheffer and Schachtschabel 2002, p. 41), both extracts ($EC_d$ and $EC_5$) are usually saturated with gypsum for soil samples containing gypsum. The solution saturated with gypsum has an EC value of 2.13 mS $cm^{-1}$. This means that if

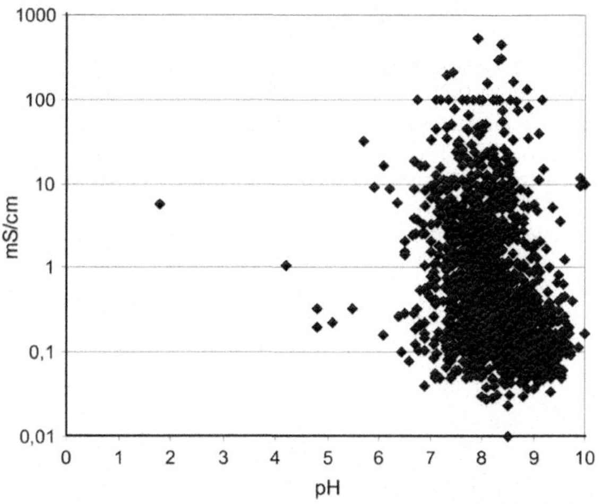


**Fig. 7.8** Electrical conductivity of 1:2.5 soil:water extract and pH values of the studied soils. pH values < 5 is from a Gleysol

the EC values of both extracts are 2.13 mS $cm^{-1}$ then the quotient $EC_d$:$EC_5 = 1$ and the dissolved salt must be gypsum. In the case of a mixture of gypsum and soluble salts, the quotient will be between 1 and 2. In the absence of gypsum, the quotient will be 2.0. Figure 7.10 shows that the $EC_d$:$EC_5$ quotient of soil samples from Central Sahara is predominantly between 1 and 2, and only sulphate-rich samples have a quotient of 1 and an $EC_5$ value of 2 mS $cm^{-1}$ (Alaily 1993; Ruck 1989).

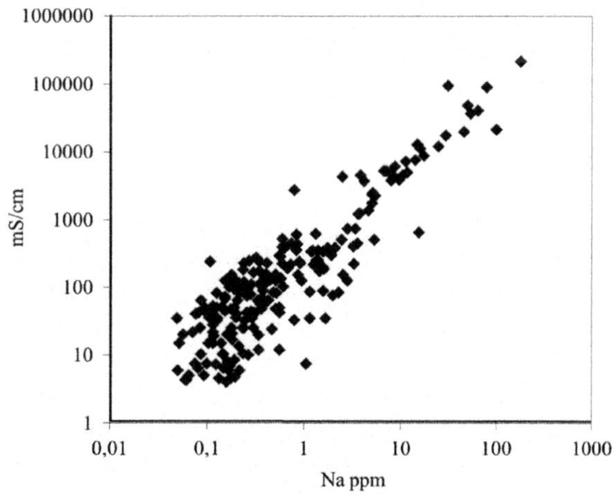


**Fig. 7.9** Relationship between water-soluble sodium and electrical conductivity of 1:2.5 soil:water extract for the investigated soils

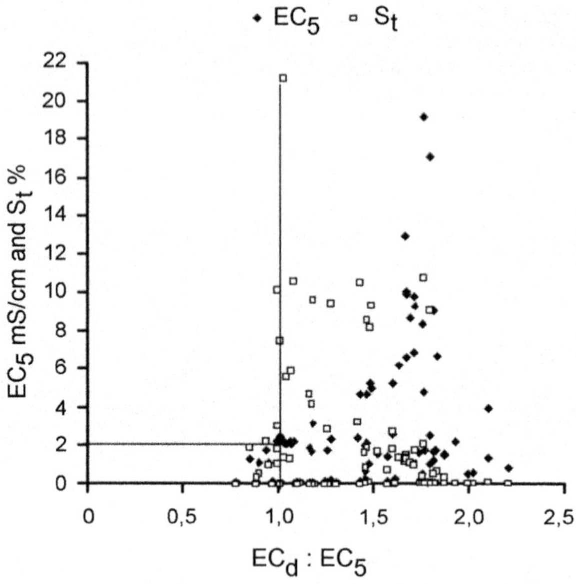


**Fig. 7.10** Relationship between $EC_5$ and total sulfur ($S_t$) on the one hand and the quotient $EC_d$:$EC_5$ of saline and gypsum-rich soil samples. EC = Electrical conductivity measured in: d = 1:2.5 and 5 = 1:5 Soil:Water extract

### *7.2.3 Desert Varnish*

Almost all the stones and coarse gravel on the soil surface in the studied region have desert varnish on their surface. The desert varnish (also called desert polish or desert patina) is a thin (2–500 μm) brown to black skin mainly consisting of iron and manganese oxides, which occurs on the surface of stones and gravel of the soil surface. Due to the desert varnish hamadas and most hills have a dark brown to black color. It is observed that the desert varnish is more intense and stronger on silicate rocks (e.g., granite or quartzite) than on limestones (Photo 7.15). The genesis of the desert varnish is still controversial. It could be produced by chemical and/or biological means. The chemical way is based on the evaporation of dew at the stone surface, whereby iron and manganese could migrate with the water from the inside of the stone to the surface where they are accumulated. This theory assumes some humidity and high evaporation for the formation of desert varnish.

The biological formation of desert varnish is based on, epi- and endo-microorganisms (chemoorganotrophic, chemolithotrophic, and phototrophic bacteria; actinomycetes; and especially fungi and lichens). These are responsible for the mobilization of iron and manganese (Bhatnagar and Bhatnagar 2005). According to Scheffer and Schachtschabel (2002, p. 526) the morning dew, which is very seldom, could be sufficient for the existence of such microorganisms. Krumbein (1993) found that in addition to iron and manganese oxides, pigments of microorganisms, especially fungi (α-carotenes, β-carotenes, and rhodoxanthene) are mixed in the composition of the desert varnish.

**Photo 7.15** Three stones with desert varnish: sandstone (**a**), quartzite (**b**) and limestone (**c**). The color of the rock can be seen at the fractures

In addition to stone type, the intensity of the desert varnish depends also on weathering duration. The formation of visible desert varnish takes a few thousand years (Photo 7.16). Thus, dateable artifacts from the Stone Age (hand axes, etc.) often have a thin desert varnish skin. It is reported that desert varnish has been formed in the Mojave Desert of the USA in less than 50 years. Moreover, desert varnish protects the stone surfaces from weathering and abrasion. Photo 7.15 shows where the desert varnish on the limestone is missing; there are clear signs of weathering.

In addition to the dark coating on the rock surfaces, the gloss on hard rocks on the soil surface is also called desert varnish or patina. This glossy coating is caused by grinding and polishing the stones with wind-sand-blower. This glossy surface is often seen on the wind-exposed sides of "windkanter", hand axes (which lie on the soil surface), and quartzite sandstone blocks.

**Photo 7.16** Ostrich eggshell shards several thousand years old with desert varnish

## 7.3 Mineralogical Composition of Clay

The results of X-ray analysis of clay from seven representative soils out of Mesozoic and Quaternary sediments are shown semi-quantitatively in Table 7.2. The clay fraction of soils from Mesozoic sediments and quaternary sediments in southwest Egypt consists mainly of kaolinite (Alaily 1987; Shadfan et al. 1985). Illite and smectite occur in very small amounts, chlorite and palygorskite are only sporadic in traces, especially in the topsoil. Pachur and Röper (1984), Smettan (1987) and Blume et al. (1987) report similar results.

Palygorskite has been detected in large quantities in tertiary sediments as well as in soils from tertiary parent material (Shadfan et al. 1985). It might occur in soils of the Egyptian plateau which are developed from tertiary sediments. This mineral belongs to the band silicates and is recognized as needles under the scanning electron microscope. In comparison, Fig. 7.11 (a) shows minerals of clay from Cretaceous sediment with well-crystallized hexagonal kaolinite as well as sharp-edged illite and (b) minerals of clay from tertiary sediment with needle-shaped palygorskite. In the case of soils developed from Mesozoic sediments, palygorskite is allochthon, imported by dusts from tertiary sediments found in north and east of the investigated region.

**Table 7.2** Mineral composition of the clay fraction from seven representative soils of the investigated area

| Tiefe | Hor | R | Kao | Chl | Ill | Sm | P |
|---|---|---|---|---|---|---|---|
| *Haplic Yermosol (126) from Quaternary fluvial sand* | | | | | | | |
| 0–23 | Aw | + | +++ | – | + | + | – |
| – 47 | Bw | + | ++++ | + | + | + | – |
| – 80 | Bwb | + | ++++ | + | + | + | – |
| – 120 | | + | ++++ | + | + | + | – |
| | Spaltenfüllung | | | | | | |
| *Haplic Yermosol (92) from Quaternary fluvial sand* | | | | | | | |
| 2–15 | Aw | + | +++ | + | + | – | + |
| – 60 | Bw | + | ++++ | + | – | + | – |
| – 120 | | + | +++ | + | - | + | - |
| Haplic Yermosol (114) from Quaternary fluvial sand | | | | | | | |
| 1–13 | Aw | + | ++++ | – | – | + | – |
| – 50 | Bw | + | ++++ | – | + | + | + |
| – 110 | C | + | ++++ | – | + | + | – |
| > 130 | | + | ++++ | + | + | + | – |
| *Lithosol (107) out of Mesozoic sandstone* | | | | | | | |
| 1–10 | Aw | + | +++ | + | + | + | – |
| > 10 | R | + | +++ | + | + | + | – |
| | Spaltenfüllung | + | +++ | – | + | + | + |
| *Orthic Solonchak (420) out of Mesozoic sandstone* | | | | | | | |
| 1–3 | Ak | + | +++ | – | + | + | + |
| – 16 | Bkz | + | +++ | + | + | + | + |
| – 41 | Bzy | + | +++ | + | + | ++ | + |
| *Orthic Solonchak (164) out of Mesozoic sandstone* | | | | | | | |
| 0–4 | Aw | + | +++ | + | + | + | – |
| – 10 | Bz | + | +++ | + | + | + | + |
| – 28 | Bz | + | +++ | + | + | + | – |
| – 40 | Bz | ++ | +++ | + | + | – | + |
| – 60 | | + | +++ | + | + | + | – |
| *Orthic Solonchak (416) out of Mesozoic silt stone* | | | | | | | |
| 0–10 | A | + | +++ | – | + | + | – |
| – 19 | Bz | + | +++ | – | + | + | – |
| – 24 | Rz | + | +++ | – | + | + | – |

*R* Residual, *Kao* Kaolinite, *Chl* Chlorite, *Ill* Illite, *Sm* Smectite, and *P* palygorskite
Means roughly: – = 0%, + = < 24%, ++ = 25–49%, +++ = 50–74%, ++++ = > 75%

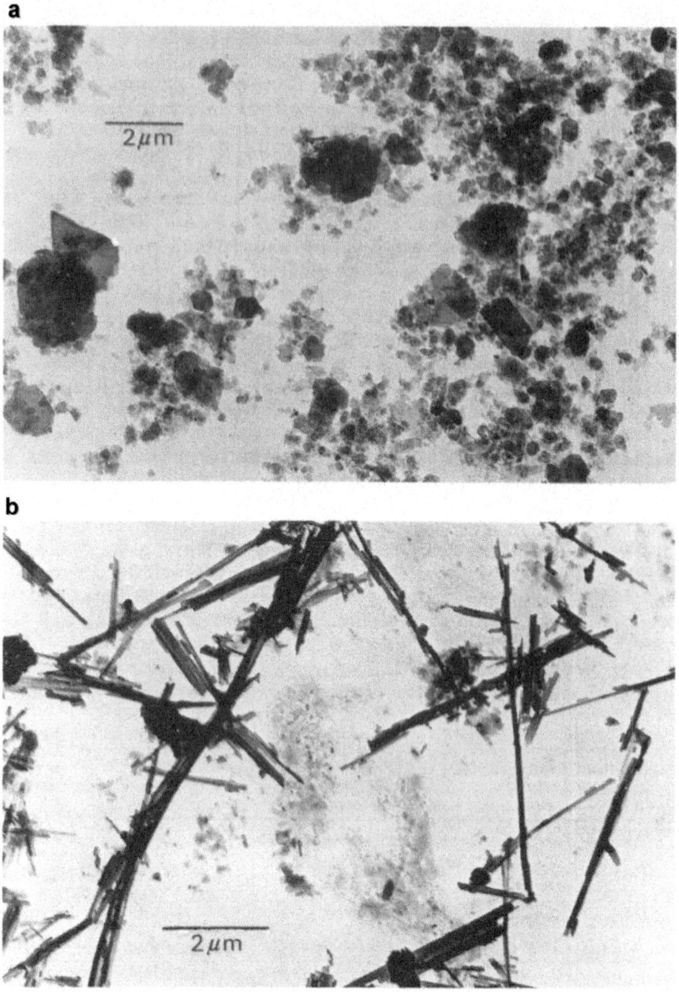


**Fig. 7.11** Scanning electron microscope images of clay (< 2 μm) from Cretaceous (**a**) and Tertiary (**b**) sediments (Shadfan et al. 1985)

Pedogenetic clay mineral conversion could not be detected in the soils studied. This was also noted by Smettan (1987). The similarity of clay mineral composition of the examined quaternary and Mesozoic sediments indicates that the origin of most quaternary sediments is from weathered product of Mesozoic rocks. The mineral composition of clay from soils out of granite, gneiss, quartzite sandstones, basalt, and playa sediments out of basalt material has not been studied. They are likely to be poor in kaolinite and rich in illite or smectite.

## References

Alaily F (1985) Nicht-hydrogene und hydrogene Boden-merkmale im Zentral-Ostsahara. Mitteilgn. Dtsch Bodenkundl Gesellsch 43/II:705–710

Alaily F (1986) Indicators of recent and subrecent soil moisture in the full arid part of the Sahara. Berliner geowiss Abh (A) 72:68–75

Alaily F (1987) Genesis of cracks in sandy soils of central east Sahara. CATENA 14:345–357

Alaily F (1993) Soil formation on limestones in the arid region of Northeast-Somalia. Catena 20:227–246

Alaily F, Blume H-P (1983) Bodengesellschaft der Vollwüste, Dakhlabecken – Ägypten. Mitteilgn Dtsch Bodenkundl Gesellsch 38:443–449

Bhatnagar A, Bhatnagar M (2005) Microbiel diversity in desert ecosystems. Curr Sci 89

Blume H-P (1985) Klimabezogene Deutung rezenter und reliktischer Eigenschaften von Wüstenböden. Geomethodica 10:91–121

Blume H-P (1986) Genesis and ecology of arid soils of the Eastern Sahara. In: Transaction XIII congress of the ISSS, Hamburg, vol III, pp 1058–1059, Göttingen

Blume H-P, Alaily F, Smettan U, Zielinski H (1984) Soil types and associations of southwest Egypt. Berliner geowiss Abh (A)50:293–302

Blume H-P, Petermann Th (1979) Yermosol-Solonchak-Bodenschaft des Fezzan, I. Genese und Ökologie. Mitteilgn Dtsch Bodenkundl Gesellsch 29:799–810

Blume H-P, Smettan U, Kalk E, Lange H (1987) Mineralbestand typischer Böden der hyperariden Wüste Süwest-Ägyptens. Berliner Geowiss Ah A78:117–126

Krumbein WE (1993) Zum Begriff Patina, seiner Beziehung zu Krusten und Verfärbungen und deren Auswirkungen auf den Zustand von Monumenten. In: Jahresberichte aus dem Forschungsprogramm Steinzerfall-Steinkonservierung (ed: Snethlage R). Ernst und Sohn, Berlin, pp 215–229

Pachur HJ, Röper P (1984) The Libyan Western Desert and Northern Sudan during the Late Pleistocene and Holocene. Berl Geow Abh A 50:249–285

Ruck A (1989) Experimentelle Ökodynamik an zwei Standorten der Extremwüste. Diss., Fachbereich 14, TU-Berlin, p 280

Scheffer F, Schachtschabel P (2002) Lehrbuch der Bodenkunde. Spektrum Akademischer VERLAG, Heidelberg und Berlin, p 593

Schlichting E, Blume H-P (1966) Bodenkundliches Praktikum. Paul Parey, Hamburg und Berlin, p 209

Shadfan H, Hussen AA, Alaily F (1985) Occurrence of palygorskite in tertiary sediments of western Egypt. Clay Minerals 20:405–413

Smettan U (1987) Typische Böden und Bodengesellschaften der Extremwüste Südwest-Ägyptens. Berliner geow Abh A 83:1–190

# Chapter 8
# Physical Weathering

**Abstract** The dominant soil-forming processes in the Central East Sahara are based on physical weathering. They lead to the break up of rocks and stones as well as to the formation of various soil structures. These processes are controlled by temperature fluctuations, frost, salt crystals, and wind abrasion (wind-sand-blower). By the first three weathering factors, the weathered product remains in situ, whereas under wind abrasion the weathered product is eroded and relocated. Insolation weathering is mainly due to temperature fluctuations. The greater the temperature fluctuation amplitude and the temperature change rate, the more intense the insolation weathering is. This also depends on the mineral composition of the stone, the color of the minerals, and the thermal conductivity of the rocks. Frost weathering occurs when sufficient moisture is present on frost days. Under recent climatic conditions, this occurs only for a few days per year. Salt weathering occurs When salts crystallize from supersaturated salt solutions, high pressures can occur if the volumes of the solution and the precipitated crystals are larger than the volume of the supersaturated starting solution. Rock fragmentation also occurs when, the volume of the reaction product in a chemical reaction, is greater than that of the reactants such as e.g. limestones containing pyrite. Wind abrasion occurs when wind speeds exceed 25 kN (12.87 m/s). Sand can be eroded and transported. The sand transported by the wind acts like sandblasters that can rub solid rock and knock out rock particles that are blown away and deposited as aeolian sediment.

**Keywords** Temperature fluctuations · Frost · Salt crystals · Wind abrasion · Wind-sand-blower

The dominant soil-forming processes in the Central East Sahara are based on physical weathering. They lead to the break up of rocks and stones as well as the formation of various soil structures. These processes are controlled by temperature fluctuations, frost, salt crystals, and wind abrasion (wind-sand-blower). By the first three weathering factors, the weathered product remains in situ, whereas under wind abrasion the weathered product is eroded and relocated. Biological weathering by plant roots has no significance in the studied region under the present climatic conditions.



F. Alaily, *Ecosystems of Extremes*, Advances in Geographical and Environmental Sciences, https://doi.org/10.1007/978-3-031-76910-8_8

## 8.1 Thermal Weathering (Insolation Weathering)

Insolation weathering is mainly due to temperature fluctuations. The greater the temperature fluctuation amplitude and the temperature change rate, the more intense the insolation weathering is. Cloud poverty and strong solar radiation cause during day time a higher heating of rocks and soil surfaces and at night to quick drop of ground temperatures than in humid regions. In 1980, Tarfawi Meteorological Station (Meteorological Station East Uweinat) measured average daily and annual temperature fluctuations of 11.4 °C and 39.2 °C respectively at 5 cm soil depth. On the soil surface, the fluctuations are likely to be much greater. Daytime temperature fluctuations of 30 and 50 °C in the top centimeters of the soil were measured by Meckelein (1959) in the Fezzan and Briem (1977) in the Tibesti.

Warming leads to expansion, cooling to contraction of the individual minerals in soil and rocks. With frequent changes between warming and cooling due to different solar radiation during day-night time (or rainfalls, which are relatively rare) and due to the different expansion and contraction of minerals, tensions occur in solid rocks which lead to its blasting. Through expansion of minerals pressure of up to 50 Mpa can arise (Scheffer and Schachtschabel 2002, p. 39). Table 8.1 shows the coefficient of linear expansion ($\alpha$ is the ratio of the relative change in length to the temperature difference) of some minerals and rocks. This shows that the various minerals and rocks have different coefficients of linear expansion, and quartz has the lowest value.

In addition, the color of minerals also plays a role, as dark minerals warm up faster and more strongly in sunlight and cool down just as quickly as light-colored ones. According to the different expansion coefficients and different colors of minerals, they will expand and contract differently. This will cause the splitting of rocks and stones. More splits will appear when temperature fluctuation is daily repeated. The enlargement of the splits is accelerated by sand or silt which trickle in the splits. The sand or silt particles prevent the splits from closing by repeated expansion. As it cools, contraction can further enlarge the split and form secondary cracks. Through this process, rocks can be crushed (Photo 8.1) and in the course of time, they break up into smaller and smaller particles. Granite breaks down more easily than basalt

**Table 8.1** Linear expansion coefficient ($\alpha \times 10^{-6}$) and thermal conductivity ($\lambda$) of some minerals and rocks as well as iron as a comparison (Kuchling 1981)

| Mineral or rock | $\alpha \times 10^{-6}$ (in 0.. 100°C) | $\lambda$ |
|---|---|---|
| Quartz | 0.45 | |
| Mica | 9.00 | 0.5–0.7 |
| NaCl | 40.00 | |
| Basalt | | 1.6 |
| Granite | 3–8 | 2.1–2.9 |
| Sandstone | 7–12 | 1.6–2.1 |
| Limestone | | 2.2 |
| Iron | 12.10 | 74.0 |

**Photo 8.1** Thermal weathering (cracking) of quarzitic sandstone

because it is coarse-grained and consists of minerals with different colors, while basalt is fine-grained and usually has a monotonous color.

Since the thermal conductivity ($\lambda$ = the amount of heat that flows through a cross-section of 1 $cm^2$ under a gradient of 1 K $cm^{-1}$ in one second, Scheffer and Schachtschabel 2002, p. 256) of rocks is low, their outer centimeters and inner parts are heated or cooled differently by rapid temperature change. They therefore expand or contract to different degrees. These lead to the formation of splits parallel to the surface and the flaking of rock plates (Photo 8.2). This type of weathering is known as shell weathering or exfoliation or desquamation and was observed on granite (Photos 8.3 and 8.4), lime stones (Photo 8.5), silt stones, and clay stones. The rock peeling of sediments occurs at laminar layer boundaries (fine stratification) and in gneiss at foliation boundaries (Photo 8.6).

## 8.2 Frost Weathering (Cryoclastic)

Frost weathering occurs when sufficient moisture is present on frost days. Under recent climatic conditions, this occurs only for a few days per year. Frost can occur at Kharga Oases in January and February and at Dakhla Oases from December to March. At the Meteorological Station of "East Uweinat" (50 km south of Tarfawi) 17 frost days were measured within 4 years (1982–1985).

The fine cracks formed by insolation weathering can be filled with water from rain or dew. By freezing, rocks can be weathered in two ways. In the first case, by

**Photo 8.2** Thermal cracking of quarzitic sandstone

**Photo 8.3** Exfoliation of granite. The granite peels off to a few cm below the surface

**Photo 8.4** Exfoliation of granite. In mm thin layers, the rock peels off

**Photo 8.5** Exfoliation of limestone

**Photo 8.6** Thermal cracking along the foliation boundaries of gneiss

freezing, water can increase its volume by about 9–10%. This results in a pressure that can be up to 2100 kg/cm$^2$ (Goudie 1995). This is much greater than the load limit of most rocks, which is generally 250 kg/cm$^2$. The second way of frost weathering is based on the property of freezing water to attract unfrozen water particles from adjacent pores. Through this process, small ice cores grow. Through the growth of ice crystals, the rock will crack. Only fine cracked rocks on the soil surface can be crushed by frost. Rocks in deeper soil layers (> 1 cm) are not affected by frost. In general, the frequency of frost is more important than the duration of the frost. Frost occurs in the investigated region only early in the morning. During the day the air temperature is well above zero, even in the coldest winters. Moreover, the number of days with frost is relatively small, and the necessary water is usually missing. Slightly more frequent frost days and slightly more humid conditions can occur in the area of Gilf Kebir. The process of frost weathering may have played a role in soil formation during the wetter periods of the Pleistocene.

## 8.3 Salt Weathering

Due to the sparse rainfall, dissolved substances are not washed out, but enriched in the soil and on the rock surfaces when the water evaporates. If salts are crystallized from oversaturated salt solutions, high pressures can occur when the volumes of the solution and the excreted crystals are greater than the volume of the oversaturated initial solution (Photo 8.7). The volume can also increase by hydrate formation, e.g.

when anhydrite (= $CaSO_4$) passes into bassanite ($CaSO_4 \cdot 1/2H_2O$) or gypsum (= $CaSO_4 \cdot 2H_2O$). In this case, pressures of several tens of MPa can occur. As with frost weathering, salt weathering is promoted by frequent dehydration and humidification.

Rock fragmentation also occurs when, the volume of the reaction product in a chemical reaction, is greater than that of the reactants such as e.g. limestones containing pyrite (Alaily 1993). The oxidation of pyrite produces sulfuric acid, which reacts with the calcium carbonate and forms calcium sulphate:

$$4FeS_2 + 15O_2 + 2H_2O \rightarrow 2Fe_2(SO_4)_3 + 2H_2SO_4.$$

$$2Fe(SO_4) + 8H_2O \rightarrow 4FeOOH + 6\ H_2SO_4.$$

$$H_2SO_4 + CaCO_3 \rightarrow CaSO_4 + CO_2 + H_2O.$$

The molecular volume of sulphate is much larger than that of carbonate; this causes the cracking of limestones containing pyrite.

Salt weathering can occur in all horizons of terrestrial soils that are saline or pyrite-rich. Compared to physical weathering by wind or temperature, salt weathering can occur both in the topsoil and in the subsoil. Whereby, it is more intense in the topsoil than in the subsoil.

**Photo 8.7** Salt cracking of siltstone. Salt crystals are visible between the plats

**Photo 8.8** "Hohlkehl" modeled by wind abrasion east of the oasis Bulaq

## 8.4 Wind Abrasion

In wind storms with wind velocity of more than 25 km (12.87 m/s), sand can be eroded and transported. The sand transported by wind acts like sandblasting blowers, which can rub solid rocks and knock out rock particles, which are blown away and deposited as Aeolian sediment. The abrasion process is most effective within the first centimeters above the ground. The particles involved in the wind abrasion jump (saltieren) on the soil surface and have a size of 0.1–0.5 mm (Scheffer and Schachtschabel 2002, p. 420). During transport, the eroded rock particles are constantly knocked against hard objects and are further crushed. In this way they get a spherical shape. With increasing height above the ground, the amount and size of hopping particles decrease and thus also the abrasion force of the wind.

Depending on the rock hardness, wind abrasion shapes different rock forms. On playas various shaped Yardangs can be formed and out of layered sediment "Hohlkehle" (Photo 8.8) and mushroom rocks are formed (Photos 8.9 and 8.10). Through sandblasting blowers, big stones (head-sized stones) on the soil surface are rubbed and smoothed in the wind direction. In this way, they are polished and they get a shiny surface. In addition, so-called "windkanter" are formed by this process. The polished surface is set against the other surfaces with a straight edge (Photo 8.11). If the stone changes its position to the main wind direction, it will be polished from a different side and several edges are then formed.

**Photo 8.9** Mushroom rock modeled by wind abrasion at the Gara Oasis in northwest Egypt

**Photo 8.10** Mushroom rock modeled by wind abrasion at Gilf Kebir

**Photo 8.11** Single-edged "windkanter" shaped by wind abrasion (sandblasting blower)

## References


Alaily F (1993) Soil formation on limestones in the arid region of Northeast-Somalia. Catena 20:227–246 (Cremlingen)

Briem E (1977) Beiträge zur Genese und Morphodynamik des ariden Formensatzes unter besonderer Berücksichtigung des Problems der Flächenbildung. Berliner Geograph Abh 26:89 (Berlin)

Goudie A (1995) Physische Geographie. Spektrum Akademischer Verlag, Berlin, pp 402

Kuchling H (1981) Taschenbuch der Physik, 3 edn, Frankfurt, p 678

Meckelein W (1959) Forschung in der Zentralen Sahara, Braunschweig, p 181

Scheffer F, Schachtschabel P (2002) Lehrbuch der Bodenkunde. Spektrum Akademischer Verlag, Heidelberg und Berlin, p 593

# Chapter 9
# Genesis of Soil Cracks

**Abstract** In the Central East Sahara, crack polygons of different sizes occur in almost all soils. Cracks controlled by wetting and drying in clay and loamy soils are called desiccation cracks or dehydration cracks (Hydro-Crack) and those in sandy soils are classified as thermal cracks (Thermo-Cracks). In sandy soils, cracks are formed through expansion and contraction of the extreme dry pedon as a result of daily and annual temperature fluctuations as well as the absence of any biological activity. Extremely dry sand particles can expand and shrink due to temperature fluctuations and arrange themselves into a higher bulk density. When a high bulk density is reached, the sand particles are pressed together by expanding and bound together by cohesive force. At last, the coherent soil mass must crack into fine fissures by shrinkage. The cracks created in this way can be filled with drifting sand and thus preserved. This means that the crack cannot close when the soil expands when heated. But it can get wider when it cools down. This causes the sand to slide deeper and more sand can trickle into the crack. As the cracks get wider, the soil surface has to bulge up to match the increase in volume. This can create a Gilgai-like micro-relief.

**Keywords** Crack polygons · Desiccation cracks · Dehydration cracks · Hydro-crack · Thermal cracks · Thermo-cracks · Temperature fluctuation · Gilgai

In the Central East Sahara, crack polygons of different sizes occur in almost all soils. The formation of such cracks is generally caused by swelling and shrinkage of the soil pedon. This is controlled by wetting and drying the soil but also by temperature fluctuations. By wetting and drying, cracks can be formed in clay and loamy soils, but not in sandy soils. In sandy soils, cracks are formed through expansion and contraction of the pedon as a result of daily and annual temperature fluctuations. Cracks controlled by wetting and drying in clay and loamy soils are called desiccation cracks or dehydration cracks (Hydro-Crack) and those in sandy soils are classified as thermal cracks (Thermo-Cracks).



F. Alaily, *Ecosystems of Extremes*, Advances in Geographical and Environmental Sciences, https://doi.org/10.1007/978-3-031-76910-8_9

## 9.1 Thermal Cracks (Thermo-Cracks)

The only possible explanation for the formation of cracks in sandy soils is the extreme dryness and continentality (hence thermo-cracks) of the area, as well as the absence of any biological activity in the soil. This is supported by several observations made in the field:

(1) In the seasonally humid semi-arid area on the Mediterranean coast in northern Egypt, such cracks in sandy substrates are not found.
(2) Some cracks reach soil depths that are hardly influenced by changes in moisture content (Photo 9.1).
3) The low clay content of the soils (< 4%), which consists mainly of kaolinite, does not indicate that the cracks were formed by swelling and shrinkage as a result of wetting and drying.
4) Cracks also occur in lithosols, not only in the weathered soil layers but also in the unweathered solid rock (R horizon) (Photo 9.2).

Based on these observations, it can be concluded that the cracks, especially in sandy soils, were not caused by wetting and drying, but by daily and annual temperature fluctuations.

This hypothesis is summarized in Fig. 9.1. The circles represent sand particles in warm and cool states. The sketch shows that extremely dry sand particles can expand and shrink due to temperature fluctuations and arrange themselves into a higher bulk density. When a high bulk density is reached, the sand particles are pressed together by expanding and bound together by cohesive force. At last, the coherent soil mass must crack into fine fissures by shrinkage. This method of cracking in sand can only occur in dry sand. In moist sand, the water menisci prevent the rearrangement of the sand (Fig. 9.2).

The cracks created in this way can be filled with drifting sand and thus preserved (Fig. 9.3). This means that the crack cannot close when the soil expands when heated. But it can get wider when it cools down. This causes the sand to slide deeper and more sand can trickle into the crack. As the cracks get wider, the soil surface has to bulge up to match the increase in volume. This can create a Gilgai-like micro-relief.

The crack width (W) caused by daily and annual temperature fluctuations was calculated by Alaily (1987). This calculation is based on the daily and annual temperature fluctuations ($\Delta$t) at 5, 10, 20, 50, and 100 cm soil depth and the assumption that the strain coefficient of the soil is similar to that of the sandstone ($\alpha = 7 \times 10^{-6}$ and $12 \times 10^{-6}$, Kuchling 1981). Using the parameters mentioned, the daily and annual shrinkage amounts for a 1 m long soil monolith (l) were calculated using the following formula:

$$W = l\alpha\Delta t$$

The results of the calculation are shown in Table 9.1. From the table, it can be seen that temperature fluctuations can cause cracks with widths that are larger than the diameter of fine sand. Such cracks can be filled with fine sand and silt.

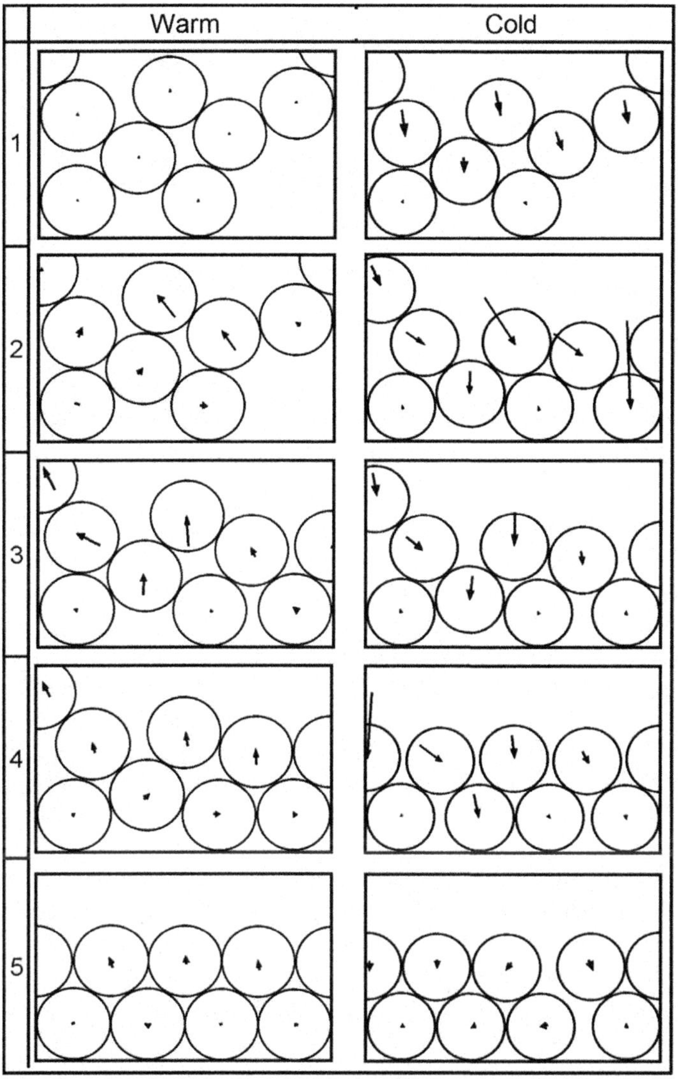


**Fig. 9.1** Rearrangement of sand particles by stretching and contraction, leading to soil compaction (5th row left) and formation of cracks (5th row right). The arrows show the extent of the displacement and the direction

From the data in Table 9.1 it can be seen;

1. The crack width or the sum of all crack widths decreases with depth.
2. The amount of expansion and contraction in the top few centimeters is 3.5 times greater than at a depth of 5 cm; this leads to the formation of a platy structure in the topsoil.

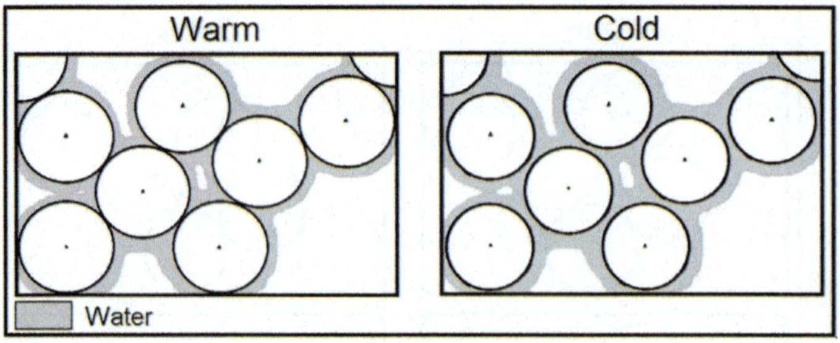


**Fig. 9.2** Water menisci prevent the rearrangement of sand particles through expansion and contraction

3. Cracks in the subsoil (>50 cm depth) are caused by annual temperature fluctuations, while in the topsoil (<50 cm depth) they are caused by daily temperature fluctuations.

The cracks can be divided into three categories: cracks, fissures and mega-cracks. The process of crack formation especially that of mega-cracks, shapes the terrain morphology. Crack formation and the associated increase in soil volume create saddles and hollows. The undulating relief in many areas with mega-cracks (e.g., north of Bir Safsaf) can be attributed to crack formation (Photo 9.3). In addition, the relief can be indirectly reshaped by crack polygons and water erosion. For example, crack polygons on the ground surface provide a favorable channel network for surface water flow. The contents of the crack can be easily washed away by surface water. This could change the terrain morphology during episodic rain events (with abundant precipitation) (Photo 9.4).

A similar theory for the formation of cracks in sandy soils was described by Blume (1986). According to this theory, the soil must have a minimum content of easily soluble salts, which act as a binding agent between the sand particles. The initial stage of crack formation occurs through severe soil drying, and the cracks widen due to expansion and contraction of the soil as a result of temperature fluctuations.

## 9.2 Dehydration Cracks (Hydro-Crack)

Dehydration cracks (Hydro-Crack) or desiccation cracks occur when the binding forces of water menisci become smaller than the tension forces between soil-minerals. They can be observed in clay-rich soils after dehydration. They are mainly formed in arid and semi-arid areas in Vertisols. The latter are soils with more than 30% clay and have subsoil aggregates with shiny surfaces known as stress cutans or shear surfaces or "slickensides" (FAO-UNESCO 1988; ISSS ISRIC FAO 1998). They are located at the base of the cracks and are formed by mutual pressures of the aggregates by swelling as a result of water absorption. Frequent swelling and shrinkage orients (textured) the clay minerals in parallel directions, which causes the

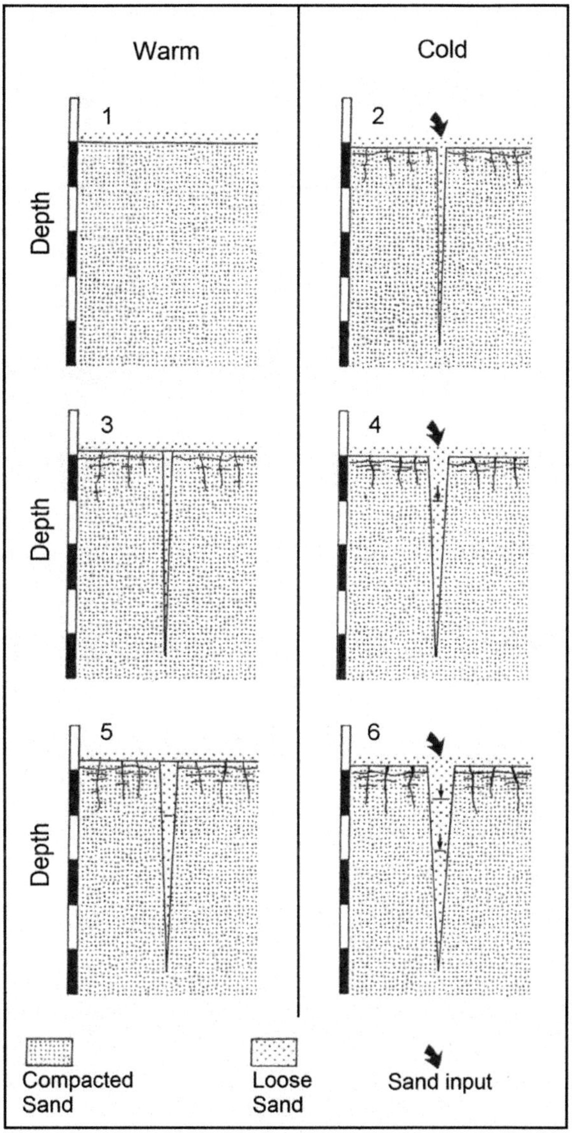


**Fig. 9.3** Stages of crack widening and soil surface uplift

gloss of the aggregate surfaces. The swelling and shrinkage processes of Vertisols are mainly controlled by the addition or decrease of the adsorption water on the soil particles or the capillary water. This depends on the grain size, while the clay mineral composition plays a minor role. Thus, Vertisols with different clay minerals can be found. Vertisols with wide dehydration cracks can be rich not only on smectite or vermiculite (Scheffer and Schachtschabel 2002) but also on kaolinite (Ahmad 1983) or palygorskite (Alaily 1996).

**Photo 9.1** Mega-crack in a cambic arenosol

**Photo 9.2** Crack in the A, B, and R horizons of a Leptosol

**Table 9.1** Daily and annual temperature fluctuations and possible sum of all crack widths (W) caused by shrinkage of 1 m of soil in the horizontal dimension

| Depth in cm | Temperature fluctuations ($\Delta t$) in °C | | Crack width ($W$) in µm | | | |
|---|---|---|---|---|---|---|
| | Daily | Annually | Daily if $\alpha =$ | | Annually if $\alpha =$ | |
| | | | $7 \times 10^{-6}$ | $12 \times 10^{-6}$ | $7 \times 10^{-6}$ | $12 \times 10^{-6}$ |
| 1[a] | 40.0 | – | 280 | 480 | – | – |
| 5[b] | 11.4 | 39.2 | 80 | 137 | 274 | 470 |
| 10[b] | 6.2 | 28.8 | 43 | 74 | 202 | 346 |
| 20[b] | 2.1 | 25.0 | 15 | 25 | 175 | 300 |
| 50[b] | 0.5 | 18.6 | 4 | 6 | 130 | 223 |
| 100[c] | 0.0 | 8.2 | 0 | 0 | 57 | 98 |

[a]Determined by Vahrson et al (1986)
[b]Calculated from the daily temperature values (1982) of the meteorological station "East Uweinat"
[c]Calculated from the monthly average temperatures of the Dakhla Meteorological Station (1964–1975)

**Photo 9.3** Undulated surface morphology caused by crack polygons (polygon diameter approx. 1.5 m)

The crack widths and depths depend primarily on the clay content and the degree of dehydration. As the clay content and degree of drying increase, the depth and width of the cracks increase. In the arid region of northern Somalia, the verisols have cracks with widths of 5 cm (in 20 cm depth) and depths of 120 cm (Alaily 1996).

**Photo 9.4** Remains of an old land surface created by soil erosion of crack polygons

There the "Slickensides" horizon is located at a depth of 92–140 cm. On the ground surface, the desiccation cracks are arranged polygonal (Photo 9.5) or square (Photo 9.6) and rarely in semicircles (Photo 9.7). The maximum width of hydro-cracks is usually much narrower than that of thermo-cracks. Thermo-cracks can be several decimeters wide, while hydro-cracks can be a few centimeters wide.

All observed desiccation cracks in the investigated region outside the populated oases were formed after the last wet climate phase. They are found in Vertisols out of clay-rich playas and Mesozoic sediments. Such Vertisols show cracks of over 1 cm width and up to 160 cm deep. "Slickensides" at a depth of 70 cm were described by Smettan (1987). Often, they were observed in very shallow depths (< 25 cm) even at the ground surface. This indicates that the Vertisols are relic and have been strongly eroded by wind (Photos 9.8 and 9.9). They do not show Gilgai micro-relief, which can be observed in recent or active Vertisols (Photo 9.10). Such micro-relief is caused by frequent soil swelling and shrinkage during the peloturbation process (Scheffer and Schachtschabel 2002). It must be assumed that the soil swelling phase is extremely rare under present climate conditions and outside the populated oases. It is likely to occur at intervals of several hundred years. In populated oases, soil swelling can take place by irrigation of clay-rich soils; desiccation cracks are then possible in fallow fields. The peloturbation process and slickensides formation could also be active there.

**Photo 9.5** Polygonal hydro-cracks (Wadi Oko, Sudan)

**Photo 9.6** Square hydro-cracks (Wadi Oko, Sudan)

**Photo 9.7** Semicircular hydro-cracks (Red Sea, Sudan)

**Photo 9.8** Relic hydro-cracks (Wadi Wassa, Gilf Kebir)

**Photo 9.9** Vertisol landscape strongly eroded by wind with relic hydro-cracks

**Photo 9.10** Vertisol landscape with Gilgai micro-relief in Darror valley north-east-Somalia

## References

Ahmad N (1983) Vertisols. In: Wilding LP, Smeck NE, Hall GF (eds) Pedogenesis and soil taxonomy, II. The soil orders. Elsevier, Amsterdam

Alaily F (1996b) Palygorskitreicher Vertisol im ariden Bereich NO-Somalias.Z. Pflanzenernähr. Bodenk, vol 159, pp 163–167

Alaily F (1987) Genesis of cracks in sandy soils of central east Sahara. Catena 14:345–357, Braunschweig

Blume HP (1986) Zur Bildung sandigefüllter Spalten in sandigen Wüstenböden. Z Pflanzenernähr Bodenk, vol 149, pp 347–349

FAO-UNESCO (1988) Soil map of the world, 1:5000 000, revised legend. Soil Res Rep 60:119 S. Rome

ISSS ISRIC FAO (1998) World reference base for soil resources. FAO, World Soul Resources Report, No. 84, Rome

Kuchling H (1981) Taschenbuch der Physik. 3 edn, Frankfurt, p 678

Scheffer F, Schachtschabel P (2002) Lehrbuch der bodenkunde. Spektrum akademischer VERLAG, Heidelberg und Berlin, p 593

Smettan U (1987) Typische Böden und Bodengesellschaften der Extremwüste Südwest-Ägyptens. Berliner geow Abh A 83:1–190 (Berlin)

Vahrson WG, Blume H-P, Alaily F (1986) Simulation of episodic rainfalls in the western desert of Egypt. Berliner geowiss Abh (A) 72:55–67 (Berlin)

# Chapter 10
# Soluble Salts, Gypsum and Carbonates

**Abstract** Due to almost lack of rainfall in Central East Sahara, salts in the soil cannot be washed out. Soluble salts are usually enriched at a small depth in soil. In addition to Solonchaks, most of the Vertisols as well as the Leptosols and all Histosols (Alaily in Geopotential and ecology—analysis of a desert region, vol 26, pp 100–123, 1993b) derived from tamarisk remains are rich in soluble salts. In well-drained sandy soils, soils in depressions, and with decreasing aridity, the salts are washed out or enriched at greater depth. Compared with the dynamic of soluble salts and gypsum which depend on their ion reaction in the solution, carbonates dynamic depends mainly on $CO_2$ partial pressure. The $CO_2$ content of soil air depends on biological activity; most soils in the area studied—outside the oases and Birs—are almost free of living organisms. Salt solubility value and the direction of water movement determine the accumulation sequence of salts in the soils. Thus, the soluble salt horizon is found above the gypsum enrichment horizon, and this above the carbonate enrichment horizon in soils that are mainly influenced by ascending water (e.g., groundwater, stagnant water). The reverse order can be found in soils in which descending water (e.g., precipitation water, irrigation water) is predominant. Enrichment of soluble salts at the soil surface from terrestrial locations does not take place, as the capillary rise of water is slower than the water loss due to evaporation, and therefore the water can only move as vapor in the dry topsoil. Lateral salt differentiation occurs in depressions that are influenced by groundwater. The highest water losses caused by evaporation are usually found in the lowest part of the depression. According to the salt concentrations and their solubility, soils at the edge of the depression are enriched mainly with carbonates and gypsum, in the center of the depression with soluble salts. This can be observed in Bir Safsaf and its surrounding area. Gypsum occurs in various forms in arid areas, but in the extreme deserts, it is usually enriched in the soil as white powder. In groundwater soils, carbonate often accumulates in concretions and soluble salts as crusts on the soil surface. Sources of salts in soils are groundwater (Phreatic salts), atmospheric dust (Atmogenic salts), and the soil parent material (Lithogenic salts). In addition, soils of inhabited oases are enriched with various salts by human activities (irrigation water and mineral fertilizers = Anthropogenic salts).



F. Alaily, *Ecosystems of Extremes*, Advances in Geographical and Environmental Sciences, https://doi.org/10.1007/978-3-031-76910-8_10

**Keywords** Solonchaks · Vertisols · Leptosols · Histosols · Tamarisk · Phreatic salts · Atmogenic salts · Lithogenic salts · Anthropogenic salts · Selenit · Marienglas · Salcrete · Gypscrete · Calcrete · Caliche · Pleistocene · Feshfesh

Soluble salts, sulphates, and carbonates can be formed and enriched in soils of arid regions in pedogenic ways. Easily soluble salts are mainly chlorides and nitrates of alkali metals as well as Na and Mg sulphates and sodium carbonate. Easily soluble salts are usually abbreviated as soluble salts. In Central Eastern Sahara, soluble salts consist mainly of Na chlorides and to a lesser extent K and Mg chlorides. In addition, Ca—carbonte, Ca—sulphate, Na— and K—nitrate can be enriched in soils of arid regions. Carbonates, sulphates, chlorides, and nitrates can be formed in the soil by biological and chemical means. The most common pedogenic minerals of salts in the investigated region are calcite ($CaCO_3$), gypsum ($CaSO_4$), and halite (NaCl).

The processes that mainly lead to calcite enrichment in soil are combined with the collective term "carbonatization", and those that lead to the enrichment of soluble salts and gypsum are summarized with the term "salinization". Nitrate formation as a result of the nitrification process only takes place biologically in the oases. This chapter covers only the chemical processes responsible for the formation of these salts.

## 10.1 Soil Salinization and Carbonatization

The ions that move more or less freely in aqueous soil solutions can cause chemical processes (ion reactions) that lead to the formation of pedogenic minerals. The characteristic of the ion reactions is that they proceed very quickly. As a result, dissolved salts can be recrystallized and enriched in the soil in a short time. This happens when the soil solution is oversaturated with salts. In warm-arid climates, the soil solution is oversaturated when water of the solution evaporates and/or the solution cools down during the night and in winter. During this process, first the sparingly soluble and finally the easily soluble salts are precipitated. Accordingly, the poorly soluble Ca carbonate will precipitate first in the soil, then the Ca sulphate, and finally the chlorides and nitrates.

Due to almost lack of rainfall in Central East Sahara, salts in the soil cannot be washed out. Soluble salts are usually enriched at a small depth in soil. In addition to Solonchaks, Histosols, Anthrosols, Leptosols, Vertisols, Fluvisols, Gypsisols, Calcisols, Cambisols, Arenosols and Regosols can contain a notable amount of soluble salts (FAO-UNESCO 1988 and ISSS, ISRIC, FAO 1998). In the central-eastern Sahara, a large proportion of the Vertisols, as well as the Leptosols and all Histosols derived from tamarisk, remains (Alaily 1993b) are rich in soluble salts.

Compared with the dynamic of soluble salts and gypsum which depend on their ion reaction in the solution, carbonates dynamic depends mainly on $CO_2$ partial pressure. The solubility of carbonates increases with increasing $CO_2$ partial pressure and vice versa:

$$(CO_3)^{2-} + CO_2 + H_2O \leftrightarrow 2(HCO_3)^-$$

When $CO_2$ escapes from the soil solution, the equilibrium will shift to the left. This happens as soil temperature increases and vice versa when the $CO_2$ content of the soil air increases. The $CO_2$ content of the soil air depends on the biological activity. In humid climates, the $CO_2$ content of the soil is about ten times higher (>0.2 vol.%) than that of the atmosphere. This $CO_2$ concentration in the soil air can only be achieved in oasis soils. Otherwise, most of the soils in the studied region are almost free of living organisms.

## 10.2 Salinization and Carbonatization Patterns and Depth Functions

The amount of enriched Allochthonous salts in terrestrial soils and the depth of the enrichment horizons depend above all on the following:

- the intensity and duration of the arid climate,
- the content of salt in the atmosphere,
- the relief type, and
- the soil-physical properties (structure, texture, and bulk density).

In well-drained sandy soils, soils in depressions, and with decreasing aridity, the salts are washed out or enriched at greater depth than in low-permeable clay-rich soils, soils on hilltop, slopes, and more arid climates. In the semi-arid region outside the coastal zone, only clay to silty soils, less often sandy soils, are enriched with salts, while in our studied region (extreme arid = precipitation 0.5 mm. $a^{-1}$) only sandy soils or soils in depressions have low-salt content, otherwise the soils are saline to very saline.

The excretion of soluble salts from the soil solution takes place mainly by water evaporation. The poorly soluble calcite (see above) is the first excreted then gypsum and finally the soluble salts. Soluble salts are also differentiated according to their solubility (Smettan 1987; Smettan and Blume 1987). Salt solubility value and the direction of water movement determine the accumulation sequence of salts in the soils. Thus, the soluble salt horizon is found above the gypsum enrichment horizon, and this above, the carbonate enrichment horizon in soils that are mainly influenced by ascending water (e.g., groundwater, stagnant water). Figure 10.1 shows the depth function of carbonate and gypsum in a relict groundwater soil out of alluvial sand located southwest of Bir Safsaf. The reverse order can be found in soils in which descending water (e.g., precipitation water, irrigation water) is predominant. Figure 10.2 shows the depth functions of carbonate, gypsum, and soluble salts in two terrestrial soils a few dozen kilometers south of Dakhla Oases.

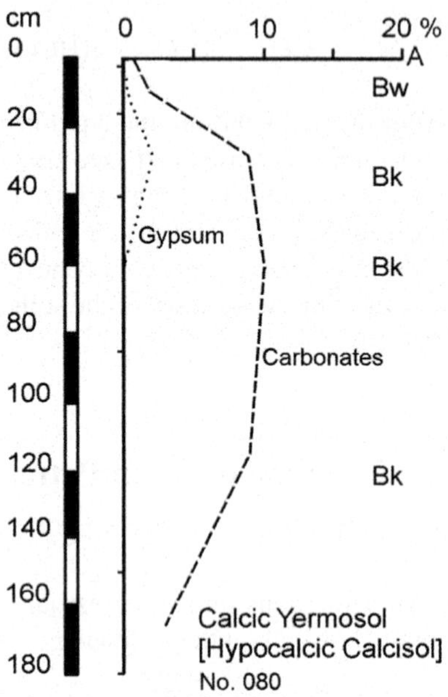


**Fig. 10.1** Carbonate and gypsum depth function in a relict groundwater soil out of alluvial sand

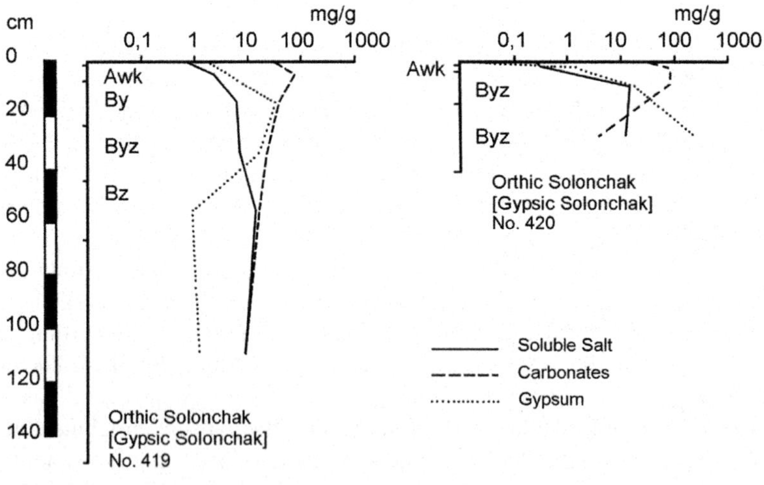


**Fig. 10.2** Depth functions of carbonate, gypsum, and soluble salts in two Orthic Solonchaks out of fluvial sediment (419) and sandstone (420) in the extreme desert

Under the extremely arid climate, carbonates are hardly leached in terrestrial soils. Figure 10.2 shows that the atmogen-enriched carbonates remained in the topsoil. The reasons for this are the very poor solubility of most carbonates, the extremely shallow infiltration of precipitation water, and the low $CO_2$ production by soil organisms. While the highest content of gypsum and soluble salt is found in the subsoil at depths between 30 and 50 cm (Smettan 1987; Ruck 1989). A sprinkling experiment on soil from alluvial sand has shown that under present climate conditions, episodic rain with a maximum of 8 mm, hardly penetrates the soil deeper than 15 cm (except in hollows and over-sanded locations); therefore, soluble salts that are enriched deeper in the soil must be relict or para-relict (Blume 1985). Enrichment of soluble salts at the soil surface from terrestrial locations does not take place, as the capillary rise of water is slower than the water loss due to evaporation, and therefore the water can only move as vapor in the dry topsoil. In addition, gypsum (or anhydrite) is often enriched on the soil surface in the center of crack polygons (Alaily 1993b). This gypsum appears as white spots directly under a silty layer (Photos 7.7 and 7.8). Even under stones on the soil surface, a pedogenous gypsum stain can often be observed (Photo 7.4).

Lateral salt differentiation occurs in depressions that are influenced by groundwater. The highest water losses caused by evaporation are usually found in the lowest part of the depression. According to the salt concentrations and their solubility, soils at the edge of the depression are enriched mainly with carbonates and gypsum, in the center of the depression with soluble salts. When groundwater reaches the soil surface, salt crusts are formed. Very hard salt crusts of 20–30 cm thickness build the topsoil from relict Gleyic Solonchaks southern Bir Safsaf (Photo 10.1). Such a hard salt crust is also referred to as salt bank or "salcrete".

In the surroundings of Bir Safsaf, the soils are strongly enriched with soluble salts and at its higher vicinity they are enriched with gypsum and on the highest part of the landscape they are strongly enriched with carbonates (Fig. 10.3). This sequence is probably caused by evaporation of Pleistocene lakes that had groundwater connections. Water evaporation from the lake must lead groundwater movement of the surrounding area in direction of the lake (or sink). In the course of increasing aridity evaporation caused the lowering of the lake water level as well as the groundwater level. The lake finally dried up. In this way, the soils in the lowest parts of the landscape were enriched with soluble salts, followed by gypsum in the higher surroundings, and carbonates in the higher parts of the area.

## 10.3 Enrichment Forms of Soluble Salts, Gypsum and Carbonates

The inorganic substances dissolved in the soil solution can accumulate in the soil in various forms. They can occur as individual minerals in different crystal sizes (μm–cm), as concretions of different sizes and shapes, or as crusts of different thicknesses.

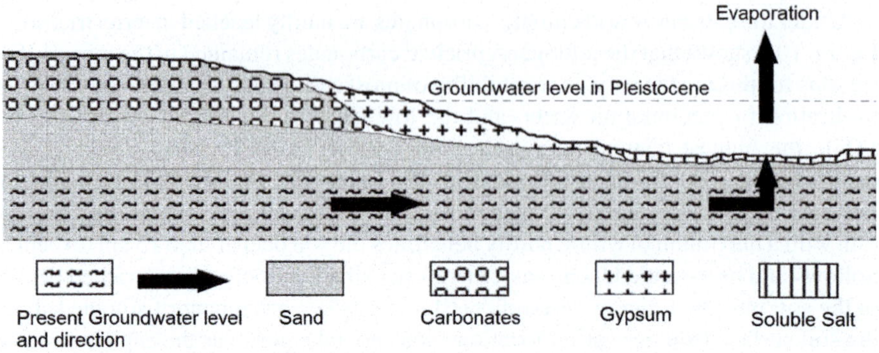


**Fig. 10.3** Schematic representation of carbonate, gypsum, and soluble salt enrichments depending on the chronology of groundwater level and the direction of groundwater movement in the area of Bir Safsaf

Soluble salts occur often as fine individual minerals and rarely in sand size. In terrestrial soils, soluble salt minerals are mainly distributed in the soil matrix. Powder-like topsoil structure (under sandy and silty soil cover) with high content of soluble salt, gypsum, and carbonates occurs in the investigated region often in terrestrial soils out of limestone and marl stones. This powder-like structure is referred to as "feshfesh" or "fluffy". The "feshfesh" topsoil horizon can be up to 10 cm thick.

If the groundwater is high, soluble salt minerals are enriched on the soil surface, where they occasionally form thin, weak crusts. Hard salt crusts (salcrete), which consist mainly of soluble salts, are found in some localities, especially in the lowlands around Bir Safsaf. These crusts can be a few centimeters to a few decimeters thick (Photos 10.14 and 10.15). They are rarely found on higher terrain positions. Salt crusts which are found on high terrain positions are remnants of old sink surfaces or dried-up lakes. They represent a relief reversal due to deflation of the surrounding area. Similar descriptions of salt crust are presented by Ganssen (1968). Slightly fragile thin salt crusts of a few mm thicknesses can be observed under tamarisk vegetation. Their formation is caused by salt excretion from the leaves of tamarisks (Photo 10.9).

Gypsum occurs in a variety of forms in arid areas (Photo 10.1), named as follows:

- White powdery crystals on aggregate surfaces or in pockets and crevices.
- Dense spheric tubers made of gypsum and soil material.
- In the form of rosettes.
- Transparent gypsum crystal referred to as Selenit or "Marienglas".
- Needle-shaped or fiber-shaped gypsum crystals.
- Hard crusts of dense gypsum (van Alphen and Rios Romero 1971), which are referred to as "Gypscrete" (Photos 10.2 and 10.3).

In the investigated region, pedogenous gypsum is often found as a snow-white powder. Needle-shaped gypsum crystals were found in the subsoil of a Chromic Vertisol (Salic Vertisol) from playa sediment in a locality east of Dakhla (10–29 cm depth). Both gypsum forms are pedogenic and were formed under the present climate conditions. Desert roses with a diameter of a few centimeters were observed on the ground surface west of Kharga. These can be from relict soils of more humid climates and they remain on the topsoil as erosion residuals. The other gypsum forms (tubers, Gypscrete, Selenite) are lithogenic and have been observed outside the investigated region in northern Egypt.

Pedogenic carbonates occur in the following forms in the studied region:

- Microcrystals of calcite well distributed in the matrix.
- Concretions with different sizes, and
- On the soil surface as a hard calcareous crust referred to as "Calcrete" or "Caliche".

The different forms of carbonate accumulation in terrestrial soils arose in the pluvial times of Pleistocene and in the humid climatic phases of the Holocene. So, they are relict pedogenic properties. Their formation depends on the water permeability as well as the water and air content of the soil. In well-drained soils, where the unsaturated water movement predominates, carbonates accumulate in fine and medium pores. In soils with inhibited water flow, fine and medium pores remain moist in the long term, so that carbonate is enriched in the aerated coarse pores (Scheffer and Schachtschabel 1992, S 381). In this way, concretions are formed (Photos 10.4 and 10.5) and become larger as the soil develops. In groundwater soils, fine and medium pores of the Go horizons are usually filled with water, so that carbonate concretions can be enriched mainly in coarse pores. In the area of Bir Sahara, Bir Tarfawi, and Bir Safsaf in the south-east of the investigated region, soils rich in carbonate concretions are common. In these soils, the carbonate concretions were formed during Pleistocene by much higher groundwater levels than at present.

North of Bir Sahara and west of Bir Tarfawi on the Selima Serir Sheet, the topsoil has in some locations calcareous crusts (Photos 10.6 and 10.7). The thickness of the calcareous crusts is on average 5–10 cm and can reach 30 cm; their carbonate content is 90% (RÖPER 1993). The formation of the calcareous crust is also a relic. Radiocarbon and U/Th dating showed an age between 18,000 and 38,000 years bp (Röper 1993 and 1988). Calcareous crusts are often referred to as "Calcrete" or "Caliche" (Reeves 1976). According to the FAO legend of the World Soil Map (1988) and the WRB (ISSS, ISRIC, FAO 1998), encrusted carbonate layers of more than 10 cm thickness are referred to as "Petrocalcic Horizon". In the hyper-arid regions of the Sahara, where average annual precipitation of < 20 mm, calcareous crusts must be formed by precipitation of carbonates in shallow ponds and later hardened by dehydration (RÖPER 1993). In addition, evaporation at sites close to groundwater table can create "phreatic calcretes" (Wright and Tucker 1991; Strong et al. 1992).

Calcareous crusts are frequently found over large areas in arid regions (Goudie 1973; EL-Aref and Abdel Wahab 1985; Buschiazzo 1990; Gauer 1991). They can form in any substrate (Blümel 1982). In carbonate-free substrates, a calcareous crust can be formed from atmospheric carbonate that is enriched with the dust in soil. Their formation requires little precipitation and high evaporation. Annual average temperatures of 16–20 °C and precipitation of 100–500 mm/a are considered suitable for the formation of calcareous crust (Goudie 1973; 1983). According to Scheffer and Schachtschabel (1992, p. 381), a consolidated calcareous horizon can be formed with a strong carbonate enrichment (of over 15% in soils rich in gravel and stones and 40% in soils free of gravel and stones). Some unconsolidated calcareous banks that appear on the soil surface due to erosion can harden into calcareous crusts.

## 10.4 Origin of Salts in Soil

Sources of salts in soils are groundwater, atmospheric dust, and soil parent material. In addition, soils of inhabited oases are enriched with various salts by human activities (irrigation water and mineral fertilizers). The amount of accumulated salts in terrestrial soils is determined by the amount of salt in the atmosphere (dust), the parent material of the soil, its water permeability, its relief position, and the duration of the arid climate conditions. In areas with high groundwater table, salt enrichment in soil also depends on the salinity of groundwater and its depth.

Depending on the sources of salt, the Solonchaks (FAO-UNESCO 1974) of Central Sahara are to be divided into four variants: atmogenic, lithogenic, biogenic, and phreatic (relict and recent). With the exception of the Phreatic Solonchak, which belongs to the Gleyic Solonchak, the other Orthics are classified as Orthic Orthic according to the FAO-UNESCO nomenclature of 1974.

### *10.4.1 Lithogenic Salts*

Lithogenic Solonchaks are found on marine limestones, marls, and siltstones. These marine sediments occur in the Tertiary and Cretaceous formations. The Dakhla, Duwi, and Mut Formations are considered to be marine deposits. They are mainly found in the northern escarpment (north of Dakhla–Kharga) and near Baris. The rocks of these formations contain fossil-soluble salt. During the process of soil formation, the soluble salt in the rocks is washed from the topsoil to the subsoil with episodic rain. As a result, the salinity in the subsoil increases. This Solonchak variant occupies large areas in northeast of the investigated region.

### *10.4.2 Atmogenic Salts*

The most common soil in southwest Egypt is the atmogenic Solonchak. The salts of this variant are mainly enriched in dry conditions with dust or sand, occasionally in dissolved state by episodic rain. For areas far from the coast, Downes (1961) stated amounts between 10 and 20 kg $NaCl.ha^{-1}.a^{-1}$. The salts of the atmosphere can be produced to a small extent by thunderstorms. As a result, nitrate can be formed in the atmosphere and fed into the soil (Mengel 1968). This explains the higher nitrate content in the Atmogenic Solonchaks compared to the Phreatic Solonchaks. In addition, the amount of salts in the atmosphere and the salt composition must be very different from region to region. Depending on the wind direction, high amounts of atmospheric salts are to be expected south and south-east of sebchas. In the region of the oases of Dakhla, Kharga, and Baris as well as at Bir Safsaf, sebchas occur over large areas.

### *10.4.3 Phreatic Salts*

In the depressions of the oasis Kharga and Dakhla, around drainage ponds as well as in areas around Bir Safsaf and Bir Tarfawi, the groundwater level is often close to the soil surface. Due to capillary rise and evaporation, the topsoil is heavily enriched with salts from groundwater. The amount of salt accumulated in the soil depends on the salt concentration of the groundwater, and the intensity and duration of the arid climate. Areas enriched with phreatic salt are called sebchas.

Typical sebcha landscapes with salt-rich Gleyic Solonchak occur in all arid regions (extreme arid, full arid, and semi-arid). The salt often forms on the soil surface a crust a few centimeters thick. At Bir Safsaf, the salt crust is several decimeters thick. It is vaulted and broken into polygons. North of Bir Tarfawi and northwest of Bir Safsaf, remnants of salt crusts can be seen in mushroom-like forms modeled by wind abrasion (Photo 10.8).

### *10.4.4 Biogenic Salts*

The biogenic salt enrichment in soil was observed under tamarisk vegetation. Due to the excretion of salt (secretion) from the tamarisk leaves and the lack of precipitation, the soils under tamatisk are strongly enriched with salts. Through the excretion of salts, the salts of the groundwater are transferred to the soil surface. The excreted salts appear as a white powder on the leaves (Photo 10.9). Table 10.1 shows the results of water analysis: from soil extract and tamarisk leaf extracts (in a ratio of 2.5:1). This shows that the latter and the soil horizon extracts have high electrical conductivity (EC) and are richer in Ca, Mg, Na and K compared to the soil extracts from the Cz horizon. In addition, the ratio of the water-soluble Na:Ca, Na:K of the tamarisk extracts and the organic soil horizon are much greater, while Na:Mg is significantly smaller than in the soil extract of the organic matter-free Cz horizon (Table 10.2, Fig. 10.4). The salt enriched in the upper soil usually forms a thin crust, which is very fragile. In some tamarisk sites, the groundwater is often at depths of over 25 m. The high salt content of the tamarisk leaves makes their microbial degradation difficult. This leads to the formation of mighty mounds from the plant remains. Tamarisk vegetation can be found around the oases of Dakhla, Kharga, and Baris as well as in the southeastern area near the Birs.

**Table 10.1** Some chemical properties of water extracts (1:2.5 sample-material: water) from the fine earth (< 2mm) and tamarisk remains

| Depth | Hor. or Sample no | Ca (mg/kg) | Mg (mg/kg) | Na (mg/kg) | K (mg/kg) | EL (mS/cm) |
|---|---|---|---|---|---|---|
| *Tamarisk remains* | | | | | | |
| | E003.00 | 1455 | 1250 | 7249 | 525 | 11.250 |
| *Biogenic Orthic Solonchak (758)* | | | | | | |
| 2–0 | OAhz | 5872 | 4250 | 17,375 | 1490 | 29.900 |
| 0–10 | Ahz | 4327 | 3000 | 12,075 | 1022 | 25.200 |
| 10–40 | Ahz | 3137 | 1750 | 8725 | 875 | 17.660 |
| 40–90 | Cz | 283 | 49 | 642 | 77 | 15.60 |

**Table 10.2** Quotients for identifying biogenic soil salinization

| Depth | Hor. or sample no | Na/Ca | Na/K | Na/Mg |
|---|---|---|---|---|
| *Tamarisken remains* | | | | |
| | E003.00 | 4.98 | 13.81 | 5.80 |
| *Biogenic Orthic Solonchak (758)* | | | | |
| 2–0 | OAhz | 2.96 | 11.66 | 4.09 |
| 0–10 | Ahz | 2.79 | 11.82 | 4.03 |
| 10–40 | Ahz | 2.78 | 9.97 | 4.99 |
| 40–90 | Cz | 2.27 | 8.34 | 13.10 |

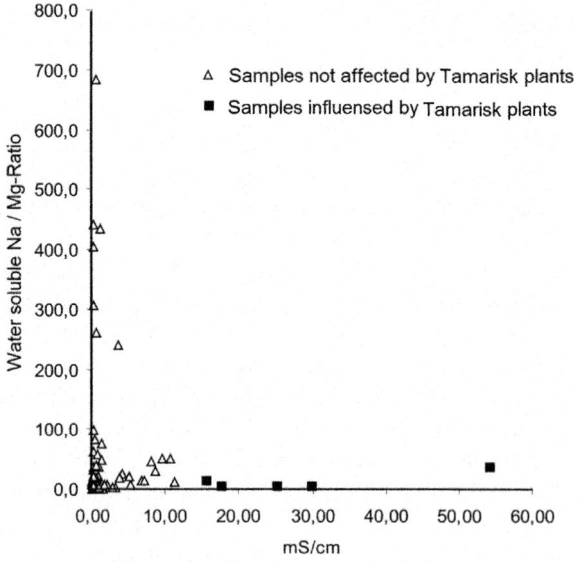


**Fig. 10.4** Relationship between the Na/Mg quotient and EC values of soil horizons from = sites free of vegetation and from soil horizons influenced by biogenic salinization

**Photo 10.1** Different forms of gypsum found in arid regions. **a**, **b** White powder, **c** rosette, **d** needle-crystals, **e** spheric tubers, **f** marienglas

### *10.4.5 Anthropogenic Salts*

The groundwater table of the oases is often raised through flood irrigation. Due to capillary rise and evaporation of the soil groundwater, salts are then enriched in the topsoil. This can be observed in soils from depressions in irrigated areas and in the vicinity of drainage ponds (evaporation ponds). In addition, the groundwater

**Photo 10.2** 25 cm thick gypsum crust over tertiary limestone (detail from Photo 10.3)

**Photo 10.3** Hard gypsum crust several decimeters thick forms the edge of a terrace

**Photo 10.4** Two calcareous concretions from the subsoil. On the soil surface, the concretions are blackened by desert varnish

**Photo 10.5** Relict groundwater soil heavily enriched with calcareous concretions, west of Bir Safsaf

**Photo 10.6** Hard calcareous crust of the topsoil from a relict groundwater soil between Black Hill and Tarfawi

is enriched with salt by fertilization. On the arable land of the oases, alkaline soils that are rich in sodium and Na carbonate are treated with gypsum in order to replace the undesired $Na^+$ on the cation-exchange-complexes with $Ca^{2+}$ cations and thus lower the pH value of the soil. This produces Na-sulphate, which is washed out by irrigation:

$$2NaHCO_3 + CaSO_4 \rightarrow CaCO_3 + Na_2SO_4 + CO_2 + H_2O$$

$$Na_2CO_3 + CaSO_4 \leftrightarrow CaCO_3 + Na_2SO_4$$

$$2Na^+ \text{ sorb} + CaSO_4 \leftrightarrow Ca^{2+} \text{ sorb} + Na_2SO_4$$

The sodium sulphate accumulates in the groundwater and accelerates the salinization of soil in depressions.

Due to minimum irrigation, the soils in the oasis areas are partly heavily enriched with salts. Soil salinization is caused by evaporation of the irrigation water and by water extraction by plants, leaving the salts of the water behind in the soil. How quickly this type of salinization occurs depends on soil-water-conductivity and the

**Photo 10.7** Calcareous crust landscape south of Bir Safsaf. It represents a dried-out wetland. The crust is blackened by desert varnish

**Photo 10.8** Salt crust remains north of Bir Tarfawi

**Photo 10.9** Salt excretion from tamarisk leaves

salt concentration of irrigation water. With decreasing soil permeability, increasing unavailable water capacity, and increasing salt concentration of the irrigation water, the risk of soil salinization increases. Extremely economical water irrigation methods (e.g., drip irrigation) also accelerate soil salinization. In the experimental farm at Bir Tarfawi, after one year of drip irrigation, the topsoils were enriched with a layer of salt (Photo 10.10). The accumulated salt comes not only from the irrigation water but also from the applied fertilizers.

In addition, various inorganic salts can be produced on arable soils of the oases by microbial degradation of organic substances (mineralization process). The most important salts formed in this way are ammonium salts, nitrates, sulphides, sulphates, and, in some places, phosphates (Alaily 1996). Ammonification is carried out by a wide range of C-heterotrophic organisms, whereas nitrification is carried out by a small number of C-autotrophic bacteria (Mengel 1985). The best-known types of bacteria responsible for the nitrification process are Nitrosomonas and Nitrobacter. Nitrosomonas converts the ammonium into nitrite ($NO_{2^-}$) and Nitrobacter converts the nitrite into nitrate. Both types of bacteria find optimal living conditions in moist, aerobic conditions and with pH values around the neutral point (pH 6–8). These conditions are met on arable land in the oases.

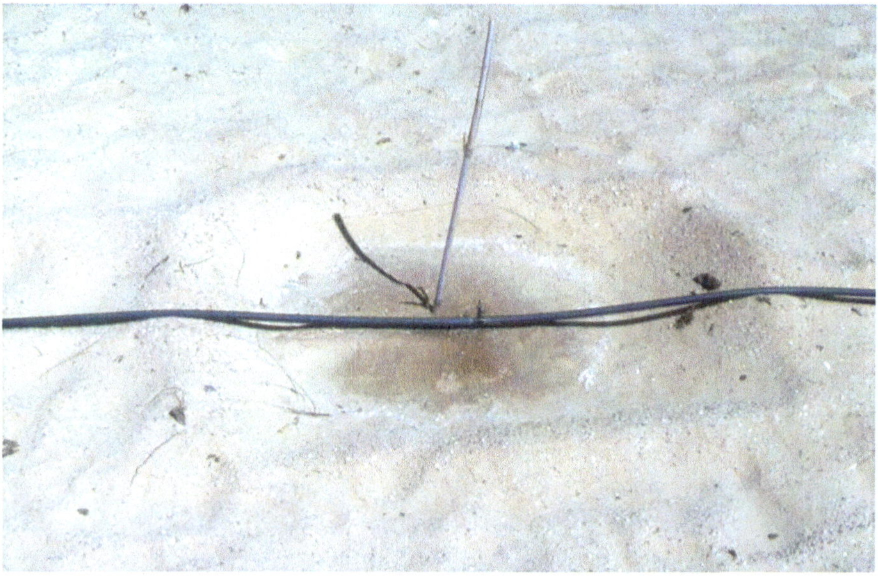

**Photo 10.10** Precipitation of white powdery salt on the soil surface due to drip irrigation of a sandy soil near Tarfawi

The salinity in the topsoil of arable land is often very heterogeneously distributed. This is due to the unevenness of soil surface. During irrigation, the slightly higher spots are usually not covered by water. There, salts are enriched through capillary rise and evaporation of water. With furrow irrigation, the highest salinity occurs on the top of the furrow and the lowest in the depression between the furrows (Wadleigh and Fireman 1949).

## 10.5 Identification of Salt-Rich Soils

Soluble salts in saline soils can be identified macroscopically and with simple chemical ways as well as their content can be estimated in the field. Their presence can also be recognized by the composition of the spontaneous vegetation or by symptoms of damage to cultivated plants (Alaily 1996). In the following chapters, the macroscopic soil properties and the electrical conductivity of saline soils are presented.

### *10.5.1 Macroscopic Characteristics of Saline Soils*

Soluble salts enriched on the soil surface (e.g., soils influenced by groundwater or stagnant water) can be easily recognized in various ways. They give the soil surface

**Photo 10.11** "Damp" soil surface due to accumulation of hygroscopic salts at the Dakhla Oasis (image area corresponds to approx. 110 × 74 cm)

characteristic properties that depend on the salt composition, soil porosity, and water tension. Due to the hygroscopic nature of soluble salts (especially $MgCl_2$-rich salt), the soil surface of such sites appears moist and oily (Photo 10.11).

Some salty soils show a peculiar micro-relief on the soil surface (Photo 10.12), which is caused by salt efflorescence. Such salt efflorescence as well as the oily soil surface can often be observed in Gleyic Solonchaks, in the vicinity of arable areas in the oases of Dakhla, Kharga, and Baris. Locations with such soil surface characteristics are often sparsely covered with plants and are therefore easy to identify on aerial photographs and satellite images (Alaily and Pohlmann 1995; Rafiq 1976).

Salt crusts of different thicknesses also occur in areas with high groundwater tables (or in areas that were influenced by groundwater in the past). The salt crusts can be a few millimeters to several decimeters thick (Photos 10.13 and 10.14). Thick salt crusts or salt crust residues can be found in areas around Birs. Thick salt crusts from relic Gleyic Solonchaks are extremely hard and can only be broken with great effort. The crust shown in Photo 10.15 could only be broken with difficulty by an excavator. Compared to gypsum crusts or calcareous crusts, salt crusts are recognizable by their taste.

Some saline soils have a so-called "Feshfesh" upper soil structure (powder-like structure = "fluffy"). The Feshfesh structure can easily be recognized by the formation of dense clouds of dust when digging or when driving over it (Photos 10.16 and 10.17). The Feshfesh structure can also be used as an indicator for a high content of

**Photo 10.12** Micro-relief caused by efflorescence of soluble salts near the Bulaq oasis

**Photo 10.13** Salt crust of a relict Gleyic Solonchak near Bir Safsaf

**Photo 10.14** A few millimeter-thin salt crust of a Gleyic Solonchak near Bulaq

**Photo 10.15** Very hard salt crust, a few decimeters thick, of a relict Gleyic Solonchak near Bir Safsaf

**Photo 10.16** The "fluffy" soil structure is easily blown into the air (e.g., by vehicles or when digging) and can form dense clouds of dust. The picture shows the dust cloud caused by our supply truck

soluble salts. Measurements of seven Feshfesh samples showed an electrical conductivity (in 1:2.5 soil: water extract) between 2.82 and 33.86 mS/cm. Such soil structures occur often in soils out of limestone and marl stones.

### *10.5.2 Electrical Conductivity of Saline Soils*

A quick method for estimating the content of dissolved salts in soil water extract or in soil solution is to measure its electrical conductivity. It was first described by Whitney and Means (1897). The method is based on the principle that the amount of electrical charge flowing through a solution increases as the salt concentration of the solution increases. In other words: the electrical resistance of a solution decreases as the salt concentration of the solution increases. Thus, the electrical conductivity is a measure of the concentration of dissolved salts in the solution. In order to determine the natural soluble salt's content of the soil solution, the electrical conductivity is usually measured in a saturated extract. The saturated extract is obtained from a soil paste which is produced by adding distilled water to the soil sample while stirring continuously until it is saturated. The water extract is then obtained by pressure or vacuum. This extraction method can only be carried out in the laboratory. Therefore, the electrical conductivity is measured in the field in a soil: water-suspension of 1:2.5 ($EC_d$:10 g soil and 25 ml water) or 1:5 ($EC_5$) and with the help of a battery-operated

**Photo 10.17** The soil material of the fluffy structure dusts and trickles out of the excavator shovel; it consists of powdery soil material mainly of very fine aggregates

electrical conductivity device. Usually, there are no suitable scales available in the field, therefore the amount of soil sample required for the suspension is determined with a letter scale or estimated with the aid of a measuring cylinder after determining the volume of 10 g of the soil sample (Schlichting et al. 1995). The values of the electrical conductivity of the saturated extract can be approximately estimated with the help of the soil texture and the water capacity of the soil as follows (Eqs. 10.1 and 10.2):

For soil samples poor in gypsum (< 0.093% sulfur or < 0.51% gypsum):

$$(250 \times EC_d) : Ws = EC_s mS\,cm^{-1} \tag{10.1}$$

For soil samples rich in gypsum (> 0.51% gypsum):

$$((250 \times (EC_d - 2.13)) : Ws)2.13 = EC_s\,mS\,cm^{-1} \tag{10.2}$$

$EC_d$ = conductivity (mS $cm^{-1}$) of 1:2.5 soil: water extract.
$EC_s$ = conductivity (mS $cm^{-1}$) of soil saturated extract.
Ws = water content of the saturated soil. This corresponds roughly to the water capacity of the soil and can be derived from the soil texture and the humus content (Schlichting et al. 1995).

2.13 = EC value of the gypsum-saturated aqueous solution at 25 °C (mS $cm^{-1}$).

**Table 10.3** The $EC_d$ equivalent values of the various soil textures for soil samples that are poor in gypsum and those rich in gypsum

| Bodenart nach | | aWC[a] | uWC[a] | WC[a] | $EC_d$-Equivalent | |
|---|---|---|---|---|---|---|
| Rothkegel (1952) | AG-Bodenkunde (1982) | Vol. % | Vol. % | Vol. % | Low-Gypsum (mS $cm^{-1}$) | High-Gypsum (mS $cm^{-1}$) |
| Sl | Sl | 18 | 8 | 26 | 1.56 | 3.47 |
| lS | Su | 20 | 6 | 26 | 1.56 | 3.47 |
| Sl | Ss | 9 | 4 | 13 | 0.78 | 2.80 |
| Sl | St | 14 | 8 | 22 | 1.32 | 3.26 |
| SL | Ls4 | 15 | 17 | 32 | 1.92 | 3.78 |
| lS | Us | 21 | 10 | 31 | 1.86 | 3.73 |
| SL | Uu | 25 | 9 | 34 | 2.04 | 3.88 |
| sL | Ut | 23 | 12 | 35 | 2.10 | 3.93 |
| lS | Lu | 16 | 18 | 34 | 2.04 | 3.88 |
| SL | Ls3 | 15 | 17 | 32 | 1.92 | 3.78 |
| LT | Ts | 12 | 31 | 43 | 2.58 | 4.34 |
| LT | Lts | 12 | 25 | 37 | 2.22 | 4.03 |
| LT | Lt | 13 | 25 | 38 | 2.28 | 4.09 |
| LT | Tl | 12 | 31 | 43 | 2.58 | 4.34 |
| T | Tt | 11 | 31 | 42 | 2.52 | 4.29 |

[a]Schlichting et al. (1995)
aWC = available water capacity, uWC = unavailable water capacity, and WC = water capacity

Soil samples with $EC_s$ of 15 mS $cm^{-1}$ are considered to be very saline. This EC value is harmful to most plants. In order to obtain a quick assessment of the salt status of soils in the field, the 15 mS/cm equivalent value of different soil textures was determined and presented in Table 10.3. The $EC_d$ equivalent corresponds to the 15 mS/cm electrical conductivity of the saturated extract. The $EC_d$ equivalent was calculated from Eqs. 10.3 and 10.4:

For soil samples poor on gypsum, the $EC_d$ equivalent is

$$= (WC \times 15) : 250 \tag{10.3}$$

For soil samples rich in gypsum, the $EC_d$ equivalent is

$$= 2.13 + ((WC \times 12.87) : 250) \tag{10.4}$$

## References


Alaily F (1993b) Soil association and land suitability maps of the western desert, SW Egypt. In: Meissner B, Wycisk P (eds) Geopotential and ecology—analysis of a desert region. Catena Supplement, vol 26, pp 100–123

Alaily F (1996) Carbonate und salze. In: Blume et al (eds) Handbuch der Bodenkunde. ecomed, Landsberg/Lech

Alaily F, Pohlmann J (1995) Soil mapping of Bir Tarfawi region (SW-Egypt) based on digital classification of landsat-MSS-data. Adv GeoEcology 28:89–107

Blume H-P (1985) Klimabezogene Deutung rezenter und reliktischer Eigenschaften von Wüstenböden. Geomethodica 10:91–121

Blümel WD (1982) Calcretes in Namibia and SE-Spain relations to substratum, soil formation and geomorphic factors. In: Yaalon H (ed) Aridic soils and geomorphic processes. Catena Suplement, vol 1, pp 67–82

Buschiazzo DE (1990) Calcrete formation in soils of the Argentinean Pampas. In: Stahr K, Blume H-P, Jahn R (eds) Gedächtniskolloquium Ernst Schlichting. Hohenheimer Arbeiten, pp 92–106

Downes RG (1961) Soil salinity in non-irrigated arable and pastoral land as the result of unbalance of the hydrologic cycle. In: Teheran symposium salinity problems in the arid zones. UNESCO, pp 105–110

EL-Aref M, Abdel Wahab S (1985) Surficial calcareous crust of caliche type along the Red Sea coast, Egypt. Geologische Rundschau 74(1):155–163

FAO-UNESCO (1974) Soil map of the world (1:5000000), vol I, Legend, Unesco Paris

FAO-UNESCO (1988) Soil map of the world, 1:5,000,000, revised legend. Soil Res Rep 60:119 S (Rome)

Ganssen R (1968) Trockengebiete. 186 S, Hochschultaschenbücher-Verlag, Mannheim, Zürich

Gauer J (1991) Bodenentwicklung und Bodengesellschaften vom Mittelmeer zur Qattara depression in Nordwestägypten. Berl geowiss Abh A 136:171 (Berlin)

Goudie AS (1973) Duricrusts in tropical and subtropical landscapes. Clarendon Press, Oxford, p 174

Goudie AS (1983) Calcrete. In: Goudi AS, Pye K (eds) Chemical sediments and geomorphology. Academic Press, London, pp 93–131

ISSS, ISRIC, FAO (1998) World reference base for soil resources. FAO, World Soul Resources Report, no 84, Rome

Mengel K (1968) Ernährung und Stoffwechsel der Pflanze, 3rd edn. VEB Gustav Fischer Verlag, Jena, p 436

Mengel K (1985) Dynamics and availability of major nutrients in soils. In: Stewart BA (ed) Advances in soil science, vol 2. Springer, New York, pp 65–131

Rafiq M (1976) Use of satellite imagery for salinity appraisal in the Indus plain. In: FAO-UNESCO: prognosis of salinity and alkalinity. Soil bulletin, vol 31, Rome, pp 141–146

Reeves CC (1976) Caliche: origin, classification, morphology and uses. Lubbock, Texas, p 233

Röper H-P (1988) Kalkkrusten in Südwest-Ägypten. Bochumer geol. U geotechn Arb 29, 163166, Bochum

Röper H-P (1993) Calcretes in the western desert of Egypt. In: Thorweihe U, Schandelmeier H (eds) Geoscientific research in NE Africa, Balkema, Rotterdam, pp 635–639

Rothkegel W (1952) Landwirtschaftliche Schätzungslehre. Ulmer, Stuttgart

Ruck A (1989) Experimentelle Ökodynamik an zwei Standorten der Extremwüste. Diss, Fachbereich 14, TU-Berlin, p 280

Scheffer F, Schachtschabel P (1992) Lehrbuch der Bodenkunde, 13 edn. Enke, Stuttgart, p 491

Schlichting E, Blume H-P, Stahr K (1995) Bodenkundliches Praktikum, 2 edn. Blackwell Wissenschafts-Verlag, Berlin, p 295

Smettan U (1987) Typische Böden und Bodengesellschaften der Extremwüste Südwest-Ägyptens. Berliner geow Abh A 83:1–190 (Berlin)

Smettan U, Blume H-P (1987) Salts in sandy desert soils Southwestern Egypt. Catena 14:333–343

Strong GE, Giles JRA, Wright VP (1992) A Holocene calcrete from North Yorkshire, England: implications for interpreting palaeoclimates using calcretes. Sedimentology 39:333–347

Van Alphen JG, de los Rios Romero F (1971) Gypsiferous soils; notes on their characteristics and management. In: International institute for land reclamation and improvement, Bulletin, vol 12, Wageningen, p 42

Wadleigh CH, Fireman M (1949) Salt distribution under furrow and basin irrigated cotton and its effect on water removal. Soil Sci Soc Amer Proc 13:527–530

Whitney M, Means TH (1897) An electrical method of determining the soluble salt content of soils. US Dept Agr Div Soils Bul 8:30 (illus)

Wright VP, Tucker ME (1991) Calcretes: introduction. Int Ass Sedimentol Reprint Ser 2:1–22

# Chapter 11
# Soil Typology

**Abstract** Depending on the parent material, terrain morphology, underground water level, and age of the soils, 13 soil units FAO-UNESCO (1974) with different "phases" were found in the investigated region. According to the WRB = "World Reference Base for soil resources" (ISSS, ISRIC, FAO in World Reference Base for Soil Resources, 1998), 54 soil units have been identified. In the WRB of 1998 compared to FAO-1974, the desert soils (Yermosols), Lithosols, and Solonchaks have been entirely redefined or redesignated. All three soil units are very common in the investigated region. Orthic Solonchaks, Lithosols, Cambic Arenosols, Haplic Yermosols, Eutric Regosols, and Calcic Yermosols are the most common soil units. Chromic Vertisols are also widespread in the oases. Outside of the irrigated areas, Vertisols are relict, as they are partly heavily eroded. Gleyic Solonchaks out of aeolian sand are found on a slope at the water hole from Bir Nakhla. In areas with high groundwater level where tamarisks stand, the extremely high dryness of the topsoil leads to the accumulation of plant residues, which form organic mounds of several meters' height (Histosol-like). These organic hills are limited to small areas at the birs. The orthic Solonchaks are divided into four variants: lithogenic Orthic Solonchak, atmogenic Orthic Solonchak, phreatic Orthic Solonchak, and biogenic Orthic Solonchak.

**Keywords** Lithosols · Cambic arenosols · Haplic yermosols · Eutric regosols · Calcic yermosols · Chromic vertisols · Lithogenic orthic solonchak · Atmogenic orthic solonchak · Phreatic orthic solonchak · Biogenic orthic solonchak · Feshfesh

The field work was carried out from 1981 to 1985 on the basis of the legend of the world soil map 1: 5,000,000 of the FAO-UNESCO (1974). Depending on the parent material, terrain morphology, underground water level, and age of the soils, 13 soil units (Table 11.1) with different "phases" were found in the investigated region. According to the WRB = "World Reference Base for soil resources" (ISSS, ISRIC, FAO 1998), 54 soil units have been identified. Table 11.1 gives the names of the soil units identified according to the FAO 1974 classification and their equivalent names according to the WRB.



F. Alaily, *Ecosystems of Extremes*, Advances in Geographical and Environmental Sciences, https://doi.org/10.1007/978-3-031-76910-8_11

**Table 11.1** Soil units determined according to the FAO classification of 1974 and their equivalent names according to the WRB (the number of soils investigated are in brackets, the names of the dominant soils are in bold and underlined, in bold and not underlined are frequent soils, in normal script are rare soils)

| FAO 1974 | WRB 1998 |
|---|---|
| Lithosol (42) | <u>**Lithic Leptosol**</u> (42) |
| Chromic Vertisol (11) | <u>**Salic Vertisol**</u> (9), Eutric Vertisol (1), Gypsic Vertisol (1) |
| Gleyic Solonchak (17) | <u>**Gleyic Solonchak**</u> (17) |
| Orthic Solonchak (189) | <u>**Gypsic Solonchak**</u> (55), **Hypersalic Solonchak** (27), **Yermic Solonchak** (19), Aridic Solonchak (15), Ochric Solonchak (11), Endosalic Gypsisol (7), Yermic Gypsisol (5), Endosalic Calcisol (6), Aridic Arenosol (4), Yermic Arenosol (4), Arenic Regosol (3), Calcic Solonchak (3), Endosalic Cambisol (3), Leptic Gypsisol (3), Leptic Regosol (3), Salic Histosol (3), Petric Calcisol (3), Aridic Regosol (2), Leptic Calcisol (2), Petrosalic Solonchak (2), Haplic Solonchak (2), Aridic Cambisol (1), Aridic Gypsisol (1), Haplic Gypsisol (1), Hyposalic Arenosol (1), Hyposalic Regosol (1), Leptic Cambisol (1), Yermic Cambisol (1) |
| Calcaric Gleysol (1) | **Calcic Gleysol** (1) |
| Eutric Gleysol (4) | **Arenic Gleysol** (4) |
| Cambic Arenosol (83) | <u>**Aridic Arenosol**</u> (21), <u>**Rubic Arenosol**</u> (21), **Arenic Regosol** (13), Hyposalic Arenosol (8), Leptic Regosol (7), Calcaric Arenosol (7), Yermic Arenosol (5), Eutric Arenosol (1) |
| Calcaric Regosol (1) | Leptic Regosol (1) |
| Eutric Regosol (13) | <u>**Arenic Regosol**</u> (6), **Leptic Regosol** (4), Aridic Leptosol (1), Eutric Leptosol (1), Eutric **Regosol** (1) |
| Takyric Yermosol (2) | Fluvic Cambisol (1), Takyric Calcisol (1) |
| Gypsic Yermosol (13) | <u>**Leptic Gypsisol**</u> (7), **Haplic Gypsisol** (2), **Aridic Gypsisol** (2), Yermic Gypsisol (1), Hyposalic Gypsisol (1) |
| Calcic Yermosol (26) | <u>**Hypocalcic Calcisol**</u> (12), **Aridic Calcisol** (4), **Leptic Calcisol** (5), Endosalic Calcisol (2), Petric Calcisol (2), Yermic Calcisol (1) |
| Haplic Yermosol (91) | <u>**Leptic Regosol**</u> (15), **Arenic Regosol** (10), **Eutric Leptosol** (9), Yermic Leptosol (8), Eutric Cambisol (6), Fluvic Cambisol (6), Yermic Cambisol (6), Calcaric Leptosol (6), Aridic Cambisol (6), Leptic Cambisol (5), Hyperskeletic Leptosol (3), Gypsiric Leptosol (3), Eutric Regosol (2), Aridic Leptosol (1), Yermic Regosol (1), Aridic Regosol (1), Endosalic Cambisol (1), Gypsiric Regosol (1), Hyposalic Regosol (1) |

In the WRB of 1998 compared to FAO-1974, the desert soils (Yermosols), Lithosols, and Solonchaks have been entirely redefined or redesignated. All three soil units are very common in the investigated region. The lithosols are no longer designated as such and are listed in the unit lithic Leptosol. The remaining Leptosols are mostly found in the Yermosols and the Eutric Regosols. In the WRB, the desert soils are also redefined on the basis of pedogenic properties; the group of Yermosols is abolished. According to the new classification, the Calcic Yermosols are divided into 6, the Gypsic Yermosol into 5, and the Haplic Yermosol into 20 new units. All Calcic Yermosols are renamed to Calcisol and all Gypsic Yermosols to Gypsisol. The Takyric Yermosol have been classified into Fluvic Cambisol and Haplic Calcisol. In the WRB of 1998, Solonchaks have been redefined: they are only those soils with a salic horizon within 50 cm depth, while in the older soil classification of 1974 Solonchaks are those soils with salic "properties" (or "high salinity") within 125 cm depth in sandy, 90 cm in loamy, and 75 cm in clay soils. Therefore, all soils that have salic properties at a depth between 50 and 125 cm are no longer to be designated as Solonchaks according to the WRB. Because of this fundamental change in the definition of the Solonchaks in the WRB version, the determined Orthic Solonchaks have been divided into 29 soil units. In addition, Cambic Arenosols with a thickness between 50 and 100 cm were assigned to the Regosols, otherwise to the Arenosols group. According to the field descriptions and the WRB, 150 Solonchaks, 73 Leptosols, 71 Regosols, 69 Arenosols, 36 Calcisols, 35 Cambisols, 33 Gypsisols, 20 Gleysols, 11 Vertisols, and 5 Histosols could be determined. The figures indicate roughly the hit frequency of the soil group mentioned.

Tables 11.2 and 11.3 list the numbers of soil units determined according to FAO-UNESCO (1974), their most common parent material and their relief position. The number of soil units determined according to the WRB, their most common parent material and relief position are given in the annex. The numbers of the respective soil unit in the tables indicate roughly their frequency. Only the Lithosols or Leptosols are underrepresented, as they can usually be seen from a distance or are located on inaccessible steep slopes or on stony terrain.

Orthic Solonchaks, Lithosols, Cambic Arenosols, Haplic Yermosols, Eutric Regosols, and Calcic Yermosols are the most common soil units. Chromic Vertisols are also widespread in the oases. Outside of the irrigated areas, Vertisols are relict, as they are partly heavily eroded, and their "slick sides" are found in the topsoil and sometimes on the soil surface. Gleyic Solonchaks out of aeolian sand are found on a slope at the water hole from Bir Nakhla. The only investigated Gleyic Solonchak out of sandstone was found at Bir Debis on a slope. In addition to the above-mentioned soil units, Calcaric Regosol (or Calcaric Fluvisol if the frequency of precipitation permits) from fluvial sediments occur in the wadis of Abu Tartur and the eastern escarpment (Smettan 1987). In areas with high groundwater level where tamarisks stand, the extremely high dryness of the topsoil leads to the accumulation of plant residues, which form organic mounds of several meters' height (Histosol-like). These organic hills are limited to small areas at the birs.

The occurrence and distribution of the various soil units can be seen on the soil association map of southwest Egypt with a scale of 1: 1,000,000 (Alaily 1993). In

**Table 11.2** Number of soil developed from hard rocks and their relief positions (P = plain or hump, S = slope, D = depression)

| | Sandstone | | | Shale | | | Limestone | | | Granite | | | Quartzite | | | Basalt | | |
|---|---|---|---|---|---|---|---|---|---|---|---|---|---|---|---|---|---|---|
| | P | S | D | P | S | D | P | S | D | P | S | D | P | S | D | P | S | D |
| Calcic Yermosol | 2 | | | | | | | | | | | | | | | 1 | | 2 |
| Gypsic Yermosol | 2 | 2 | | | | 1 | | | | | | | | | | | | |
| Haplic Yermosol | 20 | 2 | 5 | 3 | 1 | 1 | 1 | | | 1 | | 1 | 1 | | | 2 | 1 | |
| Cambic Arenosol | 1 | | | | | | | | | | | | | | | | | |
| Calcaric Regosol | | | | | | | | | | | | | | | | | | |
| Eutric Regosol | | | | | | | | | | | | | | | | | | |
| Eutric Fluvisol | | | | | | | | | | | | | | | | | | |
| Lithosol | 17 | 2 | 2 | 11 | 2 | 1 | 1 | 1 | | | | | 1 | | | | | 1 |
| Orthic Solonchak | 27 | 3 | 6 | 17 | 6 | 5 | 1 | | | 5 | 1 | 3 | | | | 1 | 1 | 1 |
| Gleyic Solonchak | 1 | | | 2 | 1 | | | | | | | | | | | | | |
| Eutric Gleysol | | | | | | | | | | | | | | | | | | |
| Chromic Vertisol | | | | | | 1 | | | | | | | | | | | | |
| Takyric Yermosol | | | | | | | | | | | | | | | | | | |

**Table 11.3** Number of soil units from Quaternary sediments their different relief positions (P = plain or hump, S = slope, D = depression)

| | Alluvial sand | | | Aeolian sand | | | Playa sediment | | | Fluvial sediment | | | Debris |
|---|---|---|---|---|---|---|---|---|---|---|---|---|---|
| | P | S | D | P | S | D | P | S | D | P | S | D | S |
| Calcic Yermosol | 8 | | 2 | 4 | | 1 | 1 | | 1 | 4 | | 2 | |
| Gypsic Yermosol | | | | | 1 | 1 | | | | 1 | 1 | 2 | 1 |
| Haplic Yermosol | 7 | | 3 | | | | 10 | | 12 | 7 | 1 | 4 | 1 |
| Cambic Arenosol | 17 | | 4 | 14 | 9 | 3 | 3 | | 3 | 17 | 6 | 8 | 1 |
| Calcaric Regosol | | | | 1 | | | | | | | | | |
| Eutric Regosol | 1 | | | 3 | 1 | | | | | 2 | | 3 | 1 |
| Orthic Solonchak | 22 | 2 | 2 | 10 | | 5 | 16 | 2 | 9 | 24 | | 11 | 14 |
| Gleyic Solonchak | 2 | | 1 | 4 | 1 | 5 | 3 | | 1 | | | | |
| Eutric Gleysol | | | | | | 1 | | | | | | 3 | 1 |
| Chromic Vertisol | | | | | | | 8 | | 1 | 1 | | | |
| Takyric Yermosol | | | | | | | 1 | | 1 | | | | |

[a]Aeolian sand includes all sandy sediments deposited by wind

[b]Fluvial sediments include recent deposits in water channels of wadis and foreland of escarpments as well as older river terraces

addition, in the work of Smettan (1987) the soil pattern for an area east of the Bulaq oasis and a second area south of Bir Tarfawi are shown on two large-scale maps.

The morphological and chemical characteristics of the common soil units (according to FAO-UNESCO 1974) are described below. The soil properties are presented in tables and in the following sequence: field color (mostly dry), soil type, skeleton in%, structure, aggregate stability, bulk density, carbonate content, root proportions, and the calculated electrical conductivity of the saturated extract (ECs). The designation of the soil unit according to the WRB (ISSS, ISRIC, FAO 1998) is given in square brackets.

## 11.1 Lithosol (I)

Lithosols are poorly developed, shallow soils with a thickness less than 10 cm and predominantly on solid rock. The Lithosols are located on Tertiary and Cretaceous sediments as well as crystalline rocks. The Tertiary and Cretaceous sediments consist mainly of sandstones, siltstones, and limestones. The crystalline rocks consist of granite, gneiss, quarzitic sandstones, and basalt. Lithosols are mainly found on steep slopes and on plains that are heavily eroded by wind (deflation plains). According to their definition, they are "lithic", "petric", and mostly "stony". Some Lithosols consist of aeolian sand (<10 cm thick) over solid rock (Photos 11.1 and 11.2) and therefore do not have a stony phase. On limestone and siltstones they are generally "saline" to very "saline" and often gypsic. Table 11.4 shows the various "phase" combinations of the Lithosols. Since the thickness of the soil is less than 10 cm, all Lithosols belong to the lithic Leptosol according to the WRB. The field properties of a representative Lithosol are shown in Table 11.5. It is found on a weak elevation in a deflation plain. The soil unit after the WRB is written in square brackets [].

**Photo 11.1** Lithosol [lithic Leptosol] from aeolian sand over sandstone

**Photo 11.2** Lithosol [Lithic Leptosol] out of sandstone with stony phase

**Table 11.4** Lithosol variants of southwest Egypt [Lithic Leptosol]

| n | lith | petri | stony | salin | ptrsl | ptrcl | ptrgp | phrti | gyp |
|---|---|---|---|---|---|---|---|---|---|
| 21 | = | + | + | | | | | | |
| 8 | = | + | + | + | | | | | |
| 8 | = | + | + | + | | | | | + |
| 4 | = | + | + | | | | | | + |
| 4 | = | + | | + | | | | | |
| 2 | = | + | | | | | | | |
| 2 | = | + | | + | | | | | + |
| 1 | = | + | | | | | | | + |
| 50 | 50 | 50 | 41 | 22 | 0 | 0 | 0 | 0 | 15 |

(lith: lithic, petri: petric, salin: saline, ptrcl: petrocalcic, ptrgp: petrogypsic, phrti: phreatic, gyp: gypsic)

**Table 11.5** Field properties of a Lithosol (No. 824) out of sandstone on a deflation plain [Lithic Leptosol]

| Depth (cm) | Horizon | Field properties |
|---|---|---|
| 0–1 | D | 8YR7/6, S, 39% Stones, single grain, very loose, contains some carbonate, $EC_s = 14$ |
| 1–6 | (A) | 7.5YR7/6, SL, 36% Stones, platy, loose, foamy, rich in carbonate, $EC_s = 4$ |
| 6–20 | R | 7.5YR6/6, Sl, 80% Stones, coherent, contains some carbonate, $EC_s = 13$ |

## 11.2 Solonchak (Z)

Solonchaks are soils with "high salinity" and have no diagnostic horizons other than A, H or a "cambic" B horizon as well as "calcic" or "gypsic" horizon. Only Orthic and Gleyic Solonchaks were found in the region.

### *11.2.1 Orthic Solonchak (Zo)*

The Orthic Solonchaks are very common and have no recent hydromorphic properties and no takyric structure. They occur on all sediments and rocks and in all landscape units. They are found on Quaternary deposits, on Tertiary and Cretaceous sediments as well as on crystalline rocks. This soil unit is found mainly on plains, occasionally on slopes and in depressions. The most frequent observed "phases" of the Orthic Solonchaks are listed in Table 11.6. This shows that about 50% of the examined soil profiles have a gypsic "phase". Orthic Solonchaks with lithic, stony or petric phase are often found on Cretaceous sediments and crystalline rocks. Orthic Solonchaks with very hard salt crust (petrosaline phase) are found in depressions and on flat humps. This phase could have been formed by drying up of inland lakes. The phreatic phase can only be found in the area of the oases and birs and is associated with Gleyic Solonchaks. In addition, Orthic Solonchaks, which have no "phases", often occur on Quaternary sediments.

According to the legend of the WRB from 1998, the Orthic Solonchak contains the following soil units: Gypsic Solonchak, Hypersalic Solonchak, Yermic Solonchak, Aridic Solonchak, Ochric Solonchak, Endosalic Gypsisol, Yermic Gypsisol, Endosalic Calcisol, Aridic Arenosol, Yermic Arenosol, Arenic Regosol, Calcic Solonchak, Endosalic Cambisol, Leptic Gypsisol, Leptic Regosol, Salic Histosol, Petric Calcisol, Aridic Regosol, Leptic Calcisol, Petrosalic Solonchak, Haplic Solonchak, Aridic Cambisol, Aridic Gypsisol, Haplic Gypsisol, Hyposalic Arenosol, Hyposalic Regosol, Leptic Cambisol, and Yermic Cambisol.

Depending on the genesis of the Orthic Solonchaks, they are divided into four variants: atmogenic, lithogenic, biogenic, and phreatic (relict and recent). Pure atmogenic Orthic Solonchaks are found on granite (Photo 11.3), gneiss, basalt, and quartzite sandstone as well as in the area of Gilf Kebir on sandstones. The lithogenic Orthic Solonchak are found on Dakhla Shale, Desert Rose sandstone as well as on limestone and marl stones (Photos 11.4 and 11.5). In the area of the birs (Tarfawi, El Shab, Kiseba, and others) and the oases (Dakhla, Kharga, and Baris) biogenic Orthic Solonchak on Tamarisk hills and relict phreatic Orthic Solonchak in sinks (which have no groundwater connection at < 50 cm soil depth) are found. The Orthic Solonchak in the area of Tarfawi and Bir Safsaf seem to have originated from former high groundwater levels. They have root remnants of unknown age.

**Table 11.6** Frequency of phases of the Orthic Solonchaks in southwest Egypt

| n | lith | petri | stony | salin | ptrsl | ptrcl | ptrgp | phrti | gyp |
|---|---|---|---|---|---|---|---|---|---|
| 31 | | | | = | | | | | |
| 23 | | | | = | | | | | + |
| 13 | + | + | + | = | | | | | + |
| 10 | | + | + | = | | | | | + |
| 10 | | | + | = | | | | | + |
| 9 | + | + | + | = | | | | | |
| 7 | | + | + | = | | | | | |
| 5 | + | | | = | | | | | + |
| 5 | + | | | = | | | | | |
| 5 | + | | + | = | | | | | + |
| 4 | + | | + | = | | | | | |
| 4 | + | + | | = | | | | | + |
| 4 | + | + | | = | | | | | |
| 4 | | + | | = | | | | | + |
| 4 | | | | = | | | | + | |
| 3 | | | + | = | | | | | |
| 1 | | + | | = | | | | | |
| 1 | + | | | = | + | | | | + |
| 1 | | + | | = | | | | + | |
| 1 | + | + | + | = | + | | | | |
| 145 | 51 | 54 | 62 | 145 | 2 | 0 | 0 | 5 | 75 |

(lith: lithic, petri: petric, salin: saline, ptrcl: petrocalcic, ptrgp: petrogypsic, phrti: phreatic, gyp: gypsic)

#### 11.2.1.1 Lithogenic Orthic Solonchak

Some characteristics of a representative lithogenic Orthic Solonchak are summarized in Tables 11.7 and 11.8. Characteristic of the lithogenic Orthic Solonchaks is carbonate enrichment in the top soil and salt as well as gypsum in the subsoil. The subsoil contains also some nitrogen. This indicates that nitrogen is in an inorganic compound. The lithogenic Orthic Solonchak often have "feshfesh" topsoil.

#### 11.2.1.2 Atmogenic Orthic Solonchak

Field properties and some analytical data of a representative atmogenic Orthic Solonchak are listed in Tables 11.9 and 11.10. Similar to the lithogenic variant, the atmogenic Orthic Solonchak is characterized by carbonate enrichment in the topsoil and

**Photo 11.3** Atmogenic Orthic Solonchak out of granite

salt as well as gypsum enrichment in the subsoil. The low value of C/N ratio indicates that nitrogen is present in an inorganic salt.

#### 11.2.1.3 Phreatic Orthic Solonchak

In contrast to the atmogenic and lithogenic Orthic Solonchak, the topsoil of the phreatic variant is enriched with salts and sometimes gypsum and carbonate (it represents relict Gleyic Solonchak). The subsoil is mostly poor in salt and gypsum, but it could be enriched with carbonates (Photo 11.6). The carbonates are sometimes found as centimeter-sized concretions. In addition, the salt enrichment horizon in the phreatic Orthic Solonchak is practically free of nitrogen. The topsoil can have extremely hard salt crust (= petrosaline) a few millimeters to a few decimeters thick or numerous centimeter-sized carbonate concretions. Often there are no rust mottles in the subsoil, which is occasionally fresh to moist. The characteristics of a phreatic Orthic Solonchak are shown in Tables 11.11 and 11.12.

**Photo 11.4** Lithogenic Orthic Solonchak [Eutric Leptosol] from Desert Rose sandstone with gypsum enrichment in the subsoil

**Photo 11.5** Lithogenic Orthic Solonchak from marl stone of the Dakhla Formation (Profile No. 416)

**Table 11.7** Field properties of a lithogenic Orthic Solonchak [Hypersalic Solonchak] (No. 416)

| Depth (cm) | Horizon | Field properties |
|---|---|---|
| 0–0.3 | A | 7.5YR7/4, sL, 27% Stones, loose, plattig und fluffy, sehr schwach, enthält wenige Gipskonkretionen, reich an Carbonat, $EL_s = 7$ mS/cm |
| 0.3–10 | Ay | 7.5YR6/2. L. 14% Steine. loose. Gypsum Cutane. crumbly. very weak. contains some gypsum concretions. very rich in carbonate. $EC_s = 7$ mS/cm |
| 10–19 | Bwz | 7.5YR6/3, L, 29% Stones, very loose, crumbly to polyhedral, very weak, contains few gypsum concretions, very rich in carbonate, $EC_s = 26$ mS/cm |
| 19–24 | Bwz | 2.5Y6/2, L, 9% Stones, very loose, gypsum cutane, polyhedral, weak, very rich in carbonate, $EC_s = 64$ |
| 24–40 | Rz | 2.5Y7/1, Sl, 96% Stones, high bulk density, very rich in carbonate, $EC_s =$ 468 mS/cm |

**Table 11.8** Some chemical properties of the fine earth of a lithogenic Orthic Solonchak [Hypersalic Solonchak] from marl

| No. | Depth (cm) | Hor | $EC_d$ (mS/cm) | Org (%) | $S_t$ (‰) | $N_t$ (ppm) | $CaCO_3$ (%[a]) | C/N |
|---|---|---|---|---|---|---|---|---|
| 416.01 | 0–0.3 | A | 0.794 | 0.18 | 0.231 | 0 | 17.1 | |
| 416.02 | 0.3–10 | Ay | 2.820 | 0.00 | 9.088 | 40 | 13.8 | |
| 416.03 | 10–19 | Bwz | 5.180 | 0.00 | 8.968 | 90 | 11.9 | |
| 416.04 | 19–24 | Bwz | 8.740 | 0.00 | 6.482 | 152 | 31.8 | |
| 416.05 | 24–40 | Rz | 10.150 | | 35.640 | 20 | | |

**Table 11.9** Field characteristics of an atmogenic Orthic Solonchak [Yermic Solonchak] (No. 164) out of sandstone on the plateau of Gilf Kebir

| Depth (cm) | Horizon | Field properties |
|---|---|---|
| 0–0.5 | D | 7.5YR6/6, S, 72% stones, Single grain, $EC_s = 3$ mS/cm |
| 0.5–4 | Akw | 5YR6/6, L, 17% stones, platy, weak, foamy, contains some carbonate. $EC_s = 1$ mS/cm |
| 4–10 | Bkz | 7.5YR6/4, L, 47% stones, gypsum cutane, gritty, very weak, some carbonate concretions, rich in carbonate, $EC_s = 34$ mS/cm |
| 10–28 | Bz | 7.5YR6/5, SL, 20% stones, gypsum cutane, single grain to gritty, moderately hard, very poor in carbonate, $EC_s = 35$ mS/cm |
| 28–40 | Cz | 7.5YR6/5, lS, 45% stones, single grain to gritty, moderately hard, poor in carbonate, $EC_s = 25$ mS/cm |
| 40–60 | Cz | 7.5YR6/5, sL, 46% stones, single grain to gritty, hard, free of carbonate, $EC_s = 38$ mS/cm |

**Table 11.10** Some chemical properties of the fine earth of an atmogenic Orthic Solonchak [Yermic Solonchak] out of sandstone on the plateau of Gilf Kebir

| No. | Depth (cm) | Hor | $EC_d$ (mS/cm) | Org (%) | $S_t$ (‰) | $N_t$ (ppm) | $CaCO_3$ (%[a]) | C/N |
|---|---|---|---|---|---|---|---|---|
| 164.00 | 0–0.5 | D | 0.114 | 0.07 | 0.21 | 90 | 1.2 | 4.6 |
| 164.01 | 0.5–4 | Akw | 0.106 | 0.00 | 0.36 | 15 | 3.18 | 0.0 |
| 164.02 | 4–10 | Bkz | 4.160 | 0.06 | 1.16 | 850 | 0.8 | 0.4 |
| 164.03 | 10–28 | Bz | 3.250 | 0.19 | 0.56 | 400 | 2.58 | 2.8 |
| 164.04 | 28–40 | Cz | 1.548 | 0.13 | 0.57 | 900 | 0 | 0.8 |
| 164.05 | 40–60 | Cz | 3.850 | 0.15 | 0.22 | 1300 | 0 | 0.7 |

**Photo 11.6** Phreatic Orthic Solonchak out of alluvial sand, heavily enriched with carbonates

#### 11.2.1.4 Biogenic Orthic Solonchak

Some properties of a biogenic Orthic Solonchak are listed in Tables 11.13 and 11.14. The biogenic Orthic Solonchak is characterized by salt and organic material enrichment in the topsoil (Photo 11.7). It has a thin salt crust on the topsoil. The organic deposits on the topsoil are very dry, undecomposed, and occur in large thicknesses. The C/N ratio of the topsoil is relatively high. This indicates that the organic material is difficult to decompose. It consists of plant remains (tamarisks) as well as insect

**Table 11.11** Field characteristics of a phreatic Orthic Solonchak (No. 343) [Hypersalic Solonchak] from aeolian sand at Bir Tarfawi

| Depth (cm) | Horizon | Field properties |
|---|---|---|
| 0–6 | Dz | 7.5YR7/5, S, Single grain, contains little carbonate, dry, $EC_s = 91$ mS/cm |
| 6–10 | Az | 7.5YR7/6, S, platy, salt crust, weakly hardened, contains little carbonate, dry, $EC_s = 1362$ mS/cm |
| 10–30 | C | 10YR7/5, S, Single grain, contains little carbonate, contains some roots, humid, $EC_s = 21$ mS/cm |
| 30–50 | C | 10YR7/5, S, Single grain, contains little carbonate, contains some roots, humid, $EC_s = 10$ mS/cm |

**Table 11.12** Some chemical properties of the fine earth from a phreatic Orthic Solonchak [Hypersalic Solonchak] at Bir Tarfawi

| No. | Depth (cm) | Hor | $EC_d$ (mS/cm) | Org (%) | $S_t$ (‰) | $N_t$ (ppm) | $CaCO_3$ (%[a]) | C/N |
|---|---|---|---|---|---|---|---|---|
| 343.01 | 0–6 | Dz | 3.61 | 0.32 | 1.45 | < 80 | 1.4 | |
| 343.02 | 6–10 | Az | 54.10 | 0.22 | 15.49 | < 80 | 1.9 | |
| 343.03 | 10–30 | C | 0.84 | 0.31 | 0.59 | < 80 | 1.6 | |
| 343.04 | 30–50 | C | 0.41 | 0.19 | 0.35 | < 80 | 0.7 | |

[a]Calculated from the total Ca content

**Table 11.13** Field properties of a biogenic Orthic Solonchak [Salic Histosol] (No. 809) from aeolian sand at Bir Tarfawi

| Depth (cm) | Horizon | Field properties |
|---|---|---|
| 0–2 | Azm | 10YR7/5, S, encrusted, moderately hard, contains little carbonate, very dry, consists of 89% vegetable and. Insect remains, $EC_s = 18{,}344$ mS/cm |
| 2–17 | Cz | 10YR7/5, Sl, very loose, single grain, contains some carbonate, very dry, $EC_s = 253$ mS/cm |
| 17–32 | Cz | 10YR7/5, fS, very loose, single grain, contains little carbonates, very dry, $EC_s = 165$ mS/cm |

and scorpion skins. With the organic material, the topsoil is enriched in nitrogen and sulfur. The carbonate content is low, and it is atmogenic.

### *11.2.2 Gleyic Solonchak (Zg)*

This soil belongs to the phreatic variants of the Solonchaks. It shows hydromorphic features within the top 50 cm (FAO-UNESCO 1974). Its salt enrichment horizon was

**Table 11.14** Some chemical properties of the fine earth from a biogenic Orthic Solonchak [Salic Histosol] at Bir Tarfawi

| No. | Depth (cm) | Hor | $EC_d$ (mS/cm) | Org (%) | $S_t$ (‰) | $N_t$ (ppm) | $CaCO_3$ (%) | C/N |
|---|---|---|---|---|---|---|---|---|
| 758.01a | Overlay | Overlay | 29.90 | 5.50 | 9.89 | 799 | 1.77 | 40.0 |
| 758.01b | 0–10 | Ahz | 25.20 | 3.33 | 4.76 | 810 | 1.86 | 23.9 |
| 758.02 | 10–40 | Ahz | 17.66 | 2.47 | 4.25 | 303 | 1.52 | 47.4 |
| 758.03 | 40–90 | Cz | 1.56 | 0.05 | 0.32 | 40 | 0.83 | 7.6 |

**Photo 11.7** Biogenic Orthic Solonchak from aeolian sand and tamarisk remains

formed by capillary rise of groundwater and evaporation at the topsoil. The Gleyic Solonchaks are found in the surroundings of the oases as well as at some birs. In the vicinity of the populated oases, the Gleyic Solonchaks are largely anthropogenic due to improper irrigation of arable land and around drainage ponds. According to the legend of the world soil map of 1988, Gleyic Solonchaks (according to FAO-UNESCO 1974) with salic property in > 30 cm depth of the soil can be described as Eutric Gleysol and those with hard topsoil crusts as Lithic Leptosol. According to the WRB (ISSS, ISRIC, FAO 1998), Gleyic Solonchak with a salic horizon between 50 and 100 cm depth is referred to as Endosalic Gleysol.

The Gleyic Solonchaks occasionally have a hard salt crust, which can be a few millimeters to a few decimeters thick. Gleyic Solonchak with "petrosaline phase" are those that have a salt crust of more than 10 cm thickness. Gleyic Solonchak, which is developed from alluvial sand or aeolian sand, has a weakly pronounced Bg horizon (Photo 11.8). Excluded is sulfide-rich Gleyic Solonchak, which has a very rusty Bg horizon (Photo 11.9). This is because the sulfides present in the Br horizons are formed under oxidation conditions into sulfuric acid and iron oxides;

$$4\,FeS_2 + 15\,O_2 + 2\,H_2O \rightarrow 2\,Fe_2(SO_4)_3 + 2\,H_2SO_4$$

$$2Fe_2(SO_4)_3 + 8H_2O \rightarrow 4FeOOH + 6H_2SO_4$$

As a result, the pH of the horizon decreases very sharply. A Gr-horizon of a Gleyic Solonchak from Bir Nakhla showed a pH of 1 after drying in the laboratory, while

**Photo 11.8** Gleyic Solonchak from aeolian sand near the oasis Bulaq

a pH of 7 was measured for the same horizon in the field. The FeOOH accumulates on the surface of the plant roots as well as in the Go horizon.

The "phases" determined in the field for the Gleyic Solonchaks are found in Table 11.15. The characteristics of a Gleyic Solonchak out of aeolian sand are summarized in Tables 11.16 and 11.17. The soil profile is located northeast near the oasis Bulaq. The analysis data show that the soil is almost free of nitrogen, the topsoil is enriched with salts, and the Bg horizon is enriched with carbonates. The Bwz, Bgz, and partly Bg horizon are enriched with $Fe_o$ and $Mn_o$.

## 11.3 Cambic Arenosol (Qc)

Cambic Arenosols are slightly altered and browned soils out of sandy material of at least 50 cm thickness (Photo 11.10). The Cambic Arenosol is found on tertiary and quaternary aeolian and fluvial sandy sediments. It is common in the Serir landscapes and in the wadis of Gilf Kebir. The Cambic Arenosols described in the field are divided into the following soil units according to the WRB of 1998: Aridic Arenosol, Rubic Arenosol, Arenic Regosol, Hyposalic Arenosol, Leptic Regosol, Calcaric Arenosol, Yermic Arenosol, and Eutric Arenosol.

**Photo 11.9** Sulfide-containing Gleyic Solonchak from aeolian sand near Bir Nakhla

**Table 11.15** Number of Gleyic Solonchak variants and their "phases"

| n | lith | petri | stony | salin | ptrsl | ptrcl | ptrgp | phrti | gyp |
|---|---|---|---|---|---|---|---|---|---|
| 7 | | | | = | | | | + | |
| 1 | | | | = | + | | | | + |
| 2 | | | | = | + | | | + | |
| 1 | + | | | = | + | | | + | |
| 1 | | + | + | = | | | | + | |
| 1 | + | | + | = | + | | | + | + |
| 1 | | | | = | | + | | | |
| 1 | | | | = | | | | + | |
| 15 | 2 | 1 | 2 | 15 | 5 | 1 | 0 | 13 | 2 |

(lith: lithic, petri: petric, salin: saline, ptrcl: petrocalcic, ptrgp: petrogypsic, phrti: phreatic, gyp: gypsic)

**Table 11.16** Field characteristics of a Gleyic Solonchak [Gleyic Solonchak] (No. 733) from aeolian sand at Bulaq

| Depth (cm) | Horizon | Field properties |
|---|---|---|
| 0–7 | Awz | 10YR7/4, mS, 1% fine gravel, single grain, loose, containing carbonate, contains some roots, dry, $EC_s = 138$ mS/cm |
| 7–27 | Bwz | 10YR7/6, mS, coherent, loose, containing carbonate, rich in roots, humid, $EC_s = 161$ mS/cm |
| 27–57 | Bgz | 10YR7/4, fS, coherent, moderately dense, rust stained, rich in carbonate, rich in roots, humid, $EC_s = 33$ mS/cm |
| 57–80 | Bg | 10YR7/6, S, 1% fine gravel, coherent, moderately dense, rust stained, rich in carbonate, rich in roots, wet, $EC_s = 9$ mS/cm |
| 80 → 90 | Br | 10YR5/6, S, 1% fine gravel, coherent, moderately dense, rich in carbonate, rich in roots, wet, $EC_s = 7$ mS/cm |

**Table 11.17** Some chemical properties of the fine earth of a Gleyic Solonchak [Gleyic Solonchak] out of aeolian sand

| No | Depth (cm) | Hor | $EC_d$ (mS/cm) | $C_t$ (%) | $N_t$ (ppm) | $CaCO_3$ (%) | $Fe_o$ (ppm) | $Mn_o$ (ppm) |
|---|---|---|---|---|---|---|---|---|
| 733.01 | 0–7 | Awz | 7.200 | 1.197 | 0 | 7.4 | 95 | 50 |
| 733.02 | 7–27 | Bwz | 8.420 | 0.563 | 0 | 4.7 | 475 | 70 |
| 733.03 | 27–57 | Bgz | 0.783 | 1.169 | 0 | 9.7 | 105 | 80 |
| 733.04 | 57–80 | Bg | 0.399 | 0.908 | 0 | 7.6 | 65 | 65 |
| 733.05 | 80– > 90 | Br | 0.292 | 0.877 | 0 | 7.3 | 80 | 55 |

Cambic Arenosols are very poor in clay. Their altered horizons (Aw and Bw) have a color of 7.5–2.5 YR. The somewhat red-colored Cambic Arenosols are usually caused by the addition of clay material from the Abu Ballas formations. In contrast to the legend of the world soil map of 1974, the diagnostic cambic horizon was defined in our investigation with a thickness of over 10 cm (instead of > 25 cm). This modification was necessary because soils that lie above solid rock and whose altered horizons have thicknesses between 10 and 25 cm could not be classified according to the legend of the world soil map (FAO-UNESCO 1974).

Table 11.18 shows that most of the Cambic Arenosols do not have phases. Occasionally saline and/or stony Cambic Arenosols are found. The latter occur mainly on deflation plains, in hilly landscapes and in the wadis of Gilf Kebir. Gypsic and/or Lithic and/or Petric Cambic Arenosols are rare. They are usually found on deflation plain and hilly landscapes.

Field properties and the results of some chemical properties of a representative Cambic Arenosol are shown in Tables 11.19 and 11.20. The EC values are relatively low. Their depth function indicates that the subsoil was enriched by rain events with soluble salts. The soluble salts are mainly atmogenic. The influence of altering can be derived from the increasing $Fe_o$:$Fe_d$ ratio in the topsoil.

**Photo 11.10** Cambic Arenosol from aeolian sand with stony phase from Wadi Maftuh in Gilf Kebir

**Table 11.18** Number of The Cambic Arenosol variants and their "phases"

| n | lith | petri | stony | salin | ptrsl | ptrcl | ptrgp | phrti | gyp |
|---|---|---|---|---|---|---|---|---|---|
| 35 | | | | | | | | | |
| 8 | | | + | | | | | | |
| 8 | | | | + | | | | | |
| 3 | | | | | | | | | + |
| 2 | + | | | + | | | | | + |
| 2 | | | + | + | | | | | |
| 2 | + | | | | | | | | |
| 1 | | + | + | | | | | | |
| 1 | | | + | | | | | | + |
| 62 | 4 | 1 | 12 | 12 | 0 | 0 | 0 | 0 | 6 |

(lith: lithic, petri: petric, salin: saline, ptrcl: petrocalcic, ptrgp: petrogypsic, phrti: phreatic, gyp: gypsic)

## 11.4 Eutric Regosol (Re)

The Eutric Regosols consist of raw soil from loose sediments (Photos 11.11 and 11.12). They are found in almost all soil associations mainly on young aeolian sediments and fluvial wadi deposits that have not been flooded for decades. On alluvial sands they occur where the old (tertiary and quaternary) soils are eroded. According

**Table 11.19** Field properties of a Cambic Arenosol [Calcaric Arenosol] (No. 081) from alluvial sand

| Depth (cm) | Horizon | Field properties |
|---|---|---|
| 0–11 | D | 7.5YR7/5, fS, 9% fine gravel, fine layered, loose, contains very little carbonate, single grain, $EC_s = 1$ mS/cm |
| 11–31 | Aw | 7.5YR7/5, fS, 5% fine gravel, fine layered, loose, contains very little carbonate, platy to gritty, very weak, $EC_s = 2$ mS/cm |
| 31–53 | BwC | 7.5YR7/6, fS, 5% fine gravel, fine layered, moderately dense, contains little carbonate, gritty, very weak, $EC_s = 11$ mS/cm |
| 53–130 | C | 7.5YR6/6, fS, fine layered, moderately dense, contains little carbonate, gritty, weak, $EC_s = 2$ mS/cm |

**Table 11.20** Some chemical properties of the fine earth from a Cambic Arenosol [Calcaric Arenosol] out of alluvial sand

| No. | Depth (cm) | Hor | $EC_d$ (mS/cm) | $N_t$ (ppm) | $CaCO_3$ (%) | $Fe_o$ (‰) | $Fe_d$ (‰) | $Fe_o$:$Fe_d$ |
|---|---|---|---|---|---|---|---|---|
| 081.01 | 0–11 | D | 0.05 | 0 | 1.0 | 0.132 | 2.2 | 0.06 |
| 081.02 | 11–31 | Aw | 0.06 | 0 | 1.0 | 0.082 | 1.9 | 0.04 |
| 081.03 | 31–53 | BwC | 0.42 | 0 | 0.9 | 0.033 | 1.5 | 0.02 |

to the WRB, Eutric Regosol with a thickness of > 25 cm are to be referred to as Arenic, Leptic, and Eutric Regosol, while those with a soil thickness of 10–25 cm are to be called Aridic and Eutric Leptosol.

Table 11.21 shows that the Eutric Regosols are sometimes lithic and/or saline and rarely petric or stony. Eutric Regosol with "petric" and/or "stony phase" are mainly found in wadi beds.

In the laboratory, a Eutric Regosol (profile No. 931) out of fluvial sand and gravel was analyzed. The soil profile comes from a depression in the bed of a wadi middle run. Results of the field description and of some laboratory analyses are given in Tables 11.22 and 11.23. The results of the analyses show that the soil has a very low electrical conductivity and low content on S, N and pedogenous iron oxides (Feo), which indicates low development of this soil.

## 11.5 Yermosol (Y)

Yermosols are soils that were developed under arid climatic conditions ("aridic moisture regime"). They come in the 22th position in the hierarchical order (after Lithosols, Vertisols, Fluvisols, Solonchaks, Gleysols, Arenosols, and Regosols) in the legend of the world soil map from 1974. In the investigated region mainly Haplic, Calcic, and Gypsic Yermosol are found. The Takyric Yermosol is relatively rare. In

**Photo 11.11** Shallow Eutric Regosol out of aeolian sand with thermal crack

**Photo 11.12** Eutric Regosol from unconsolidated aeolian sand

**Table 11.21** Number of the Eutric Regosol and its "phases"

| n | lith | petri | stony | salin | ptrsl | ptrcl | ptrgp | phrti | gyp |
|---|---|---|---|---|---|---|---|---|---|
| 10 | | | | | | | | | |
| 2 | + | | | | | | | | |
| 2 | + | | | + | | | | | |
| 2 | | | | + | | | | | |
| 1 | + | | + | + | | | | | |
| 1 | | + | | + | | | | | |
| 18 | 5 | 1 | 1 | 6 | 0 | 0 | 0 | 0 | 0 |

(lith: lithic, petri: petric, salin: saline, ptrcl: petrocalcic, ptrgp: petrogypsic, phrti: phreatic, gyp: gypsic)

**Table 11.22** Field properties of a Eutric Regosol [Arenic Regosol] (No. 931) out of fluvial sand and gravel at Wadi Maftuh in Gilf Kebir

| Depth (cm) | Horizon | Field properties |
|---|---|---|
| 0–0.5 | AC | 7.5YR7/6, S, 23% Fine gravel, platy to single grain, very weak, loose, contains very little carbonate, $EC_S = 4$ mS/cm |
| 0.5–23 | C | 7.5YR7/6, S, 20% Fine gravel, platy to single grain, very weak, loose, contains some roots, $EC_S = 3$ mS/cm |
| 23–35 | C | 7.5YR7/6, S, 57% Fine gravel, single grain, moderately dense, contains some roots, $EC_S = 5$ mS/cm |
| 35–54 | C | 7.5YR7/6, S, 20% Fine gravel, single grain, moderately dense, contains some roots, $EC_S = 3$ mS/cm |
| 54–90 | bC | 5YR6/6, Sl, 3% Fine gravel, coherent, very hard, very dense, contains some roots, $El_S = 11$ mS/cm |

**Table 11.23** Some chemical properties of the fine earth from a Eutric Regosol [Arenic Regosol] out of fluvial sand and gravel at Wadi Maftuh in Gilf Kebir

| No. | Depth (cm) | Hor | $EC_d$ (mS/cm) | Org (%) | $S_t$ (‰) | $N_t$ (ppm) | $CaCO_3$ (%) | $Fe_o$ (‰) |
|---|---|---|---|---|---|---|---|---|
| 931.01 | 0–0.5 | AC | 0.105 | 0.14 | 0.04 | < 49 | 0 | 0.06 |
| 931.02 | 0.5–23 | C | 0.086 | 0.08 | 0.03 | < 49 | 0 | 0.04 |
| 931.03 | 23–35 | C | 0.080 | 0.23 | 0.05 | < 49 | 0 | 0.04 |
| 931.04 | 35–54 | C | 0.099 | 0.04 | 0.04 | < 49 | 0 | 0.14 |
| 931.05 | 54–90 | bC | 0.066 | 0.03 | 0.06 | < 49 | 0 | |

the WRB of 1998, the desert soils are redefined and are assigned as subunits of the "Major Soil Groupings" according to their pedogenic arid properties.

### 11.5.1 *Haplic Yermosol (Yh)*

With the exception of very sandy quaternary sediments (e.g., aeolian sands), the Haplic Yermosol occurs on almost all sediments and rocks (Photo 11.13). They are common on deflation plains of "Lower Clastic" and on some serir plains as well as on the plateau of Gilf Kebir and its wadis. Deviating from the legend of the world soil map from 1974, the minimum depth of the lower limit of the cambic B horizon in this work is 10 cm instead of 25 cm. Table 11.24 shows that the Haplic Yermosols are often lithic and/or petric and/or stony and/or saline and rarely gypsic. The color of the Bw horizon is between 7.5YR and 2.5YR. According to the 1998 WRB, the investigated Haplic Yermosols are distributed into the following soil units: Leptic Regosol, Arenic Regosol, Eutric Leptosol, Yermic Leptosol, Eutric Cambisol, Fluvic Cambisol, Yermic Cambisol, Calcaric Leptosol, Aridic Cambisol, Leptic Cambisol, Hyperskeletic Leptosol, Gypsiric Leptosol, Eutric Regosol, Aridic Leptosol, Yermic Regosol, Aridic Regosol, Endosalic Cambisol, Gypsiric Regosol, and Hyposalic Regosol.

The field properties of a Haplic Yermosol out of sandstone are shown in Table 11.25. The soil profile is located in a wide depression at the Black Hill area and is lying on sandstone. The analysis data in Table 11.26 show a very low content of organic material and sulfur. An influence of alteration from the quotient $Fe_o$:$Fe_t$ is not recognizable because of the heterogeneity of the parent material. Leaching and

**Photo 11.13** Haplic Yermosol from playa sediment

**Table 11.24** Number of Haplic Yermosol and its "phases"

| n | lith | petri | stony | salin | ptrsl | ptrcl | ptrgp | phrti | gyp |
|---|---|---|---|---|---|---|---|---|---|
| 15 | | | | | | | | | |
| 12 | | | | + | | | | | |
| 9 | + | | | | | | | | |
| 8 | | | + | | | | | | |
| 6 | + | + | + | | | | | | |
| 5 | + | + | | | | | | | |
| 4 | + | + | + | + | | | | | |
| 3 | + | | | + | | | | | |
| 2 | | | + | + | | | | | + |
| 2 | | | | | | | | | + |
| 1 | + | | | + | | | | | + |
| 1 | + | | + | | | | | | + |
| 1 | | + | | | | | | | |
| 1 | + | + | | | | | | | |
| 1 | + | | + | | | | | | |
| 1 | + | + | | + | | | | | |
| 72 | 32 | 18 | 22 | 23 | 0 | 0 | 0 | 0 | 6 |

(lith: lithic, petri: petric, salin: saline, ptrcl: petrocalcic, ptrgp: petrogypsic, phrti: phreatic, gyp: gypsic)

**Table 11.25** Field properties of a representative Haplic Yermosol [aridic Cambisol] (114). The soil profile shows a crack up to a depth of more than 1.40 m

| Depth (cm) | Horizon | Field properties |
|---|---|---|
| 0–0.5 | D | 7.5YR7/7, S, 21%, single grain, very loose, contains very little carbonate, contains no roots, $EC_S = 6$ mS/cm |
| 0.5–1.5 | A | 7.5YR7/6, Sl, 0%, platy, weak, very loose, contains little carbonate, contains no roots, $EC_S = 1$ mS/cm |
| 1.5–13 | A/Bw | 7.5YR7/8, Sl, 0%, platy to gritty, medium hard, loose, contains little carbonate, contains no roots, $EC_S = 1$ mS/cm |
| 13–110 | Bw | 5YR6/6, lS, 0%, gritty to coherent, hard, moderately dense to dense, contains no roots, $EC_S = 2$ mS/cm |
| 110–130 | Bw | 5YR6/6, SL, 5%, gritty to coherent, hard, dense, contains no roots, $EC_S = 10$ mS/cm |
| > 130 | R | 7.5YR6/4, 39%, coherent, hard, very dense, contains no roots |

enrichment of soluble salts in the subsoil can be derived from the EC values, which increases significantly with the depth.

**Table 11.26** Analysis results of a representative Haplic Yermosol [aridic Cambisol] out of sandstone

| No. | Depth (cm) | Hor | $EC_d$ (mS/cm) | Org (%) | $S_t$ (‰) | $Fe_o$ (‰) | $Fe_t$ (‰) | $Fe_o$:$Fe_t$ |
|---|---|---|---|---|---|---|---|---|
| 114.00 | 0–0.5 | D | 0.175 | 0.03 | 0.12 | 0.102 | 6.50 | 0.016 |
| 114.01 | 0.5–1.5 | A | 0.057 | 0.02 | 0.05 | 0.014 | 3.45 | 0.004 |
| 114.02 | 1.5–13 | A/Bw | 0.052 | 0.04 | 0.07 | 0.018 | 4.25 | 0.004 |
| 114.03a | 13–110 | Bw | 0.169 | 0.03 | 0.06 | 0.041 | 5.00 | 0.008 |
| 114.03b | 110–130 | Bw | 0.938 | 0.06 | 0.07 | 0.053 | | |
| 114.04 | > 130 | R | 1.152 | 0.05 | 0.04 | 0.061 | 11.00 | 0.006 |

### *11.5.2 Calcic Yermosol (Yk)*

The Calcic Yermosol is defined by carbonate enrichment within a depth of 125 cm (FAO-UNESCO 1974). In the investigated region, the Calcic Yermosol can be divided into: phreatic Calcic Yermosol and terrestrial Calcic Yermosol.

According to the WRB of 1998, the Calcic Yermosol is put into a "Major Soil Grouping" and referred to as Calcisols. To this soil group, the following soil units were identified in the investigated region: Hypocalcic Calcisol, Aridic Calcisol, Leptic Calcisol, Endosalic Calcisol, Petric Calcisol, and Yermic Calcisol.

## 11.6 Phreatische Calcic Yermosol

The carbonates of the phreatic Calcic Yermosol precipitated by groundwater, so the soil can be enriched with a high amount of carbonate. This Calcic Yermosol variant occurs on large areas around Bir Tarfawi (Photo 11.14) and represents the transition area between the relict Gleyic Solonchak and the soils unaffected by recent groundwater. In this area, the parent material of the soils consists of alluvial sand. There, the carbonates are enriched in the subsoil in Bkc horizon in form of fist-sized concretions. The deeper parts of the calcareous enrichment horizons are often slightly rust-mottled (Photo 11.15). This indicates that such Calcic Yermosols were influenced by high groundwater levels in the past (phreatic Calcic Yermosols). In the upper limit of the mottled calcareous accumulation horizons, fine root residues are often found. The C14 dating of the roots indicated that they are less than 100 years old. In addition, the carbonate concretions that are lying on the soil surface are coated with desert varnish. This means that the concretions had been on the soil surface for several millennia.

The soil phases of the phreatic Calcic Yermosol are shown in Table 11.27. The "petric" and "stony phases" occur frequently and are caused by calcareous concretions. Occasionally, calcareous crusts ("petrocalcic phase" = Bkm horizon) can be found on the soil surface (Photos 10.6 and 10.7. On the soil surface the calcareous crusts is blackened by desert varnish.

**Photo 11.14** Calcic Yermosol from alluvial sand near Tarfawi

Field characteristics of a phreatic Calcic Yermosol from a serir plain near Bir Tarfawi are presented in Table 11.28. The analytical data in Table 11.29 show that (as most terrestrial soils) the electrical conductivity increases with depth. Similar to phreatic Orthic Solonchak, the phreatic Calcic Yermosol is free of nitrogen. The carbonate enrichment is below the gypsum enrichment maximum (or sulfur enrichment maximum). This indicates that the precipitation of both components was from ascending groundwater. The $Fe_o$ contents decrease with depth and indicate low alteration of the soil.

## *11.6.1 Terrestrial Calcic Yermosol*

The carbonates of the terrestrial Calcic Yermosol originate from the parent material, the atmosphere or are formed in the soil from primary components ($Ca^{++}$ and $CO_2$). Pure terrestrial Calcic Yermosol can be found on granite, gneiss, sandstones, and aeolian sand in the eastern part of the investigated region as well as at Gebel Kamil and on basalt in the area of Gilf Kebir (Photo 11.16). The carbonates are enriched in the Bk or Ak horizons in powdery consistent. With the exception of terrestrial Calcic Yermosol out of aeolian sand, the soils of this variant are shallow and/or stony. The soil phases of the terrestrial Calcic Yermosol are shown in Table 11.30.

**Photo 11.15** Calcic Yermosol buried with aeolian sand. The Calcic Yermosol is from alluvial sand, and it shows relict rust mottles at a depth of 1 m and a crack more than 2 m length

**Table 11.27** Number of phreatic Calcic Yermosol variants and their "phases"

| n | lith | petri | stony | salin | ptrsl | ptrcl | ptrgp | phrti | gyp |
|---|---|---|---|---|---|---|---|---|---|
| 5 | | | | | | | | | |
| 6 | | | | + | | | | | |
| 4 | | + | | | | | | | |
| 3 | + | + | + | | | | | | |
| 2 | | | | | | + | | | |
| 2 | | + | + | + | | | | | |
| 2 | | + | | + | | | | | |
| 1 | | | + | | | | | | |
| 1 | | + | | | | + | | | |
| 1 | + | + | + | + | | | | | |
| 1 | | + | + | + | | | | | |
| 1 | + | + | + | + | | | | | + |
| 1 | + | + | + | | | + | | | |
| 30 | 6 | 16 | 10 | 13 | 0 | 4 | 0 | 0 | 1 |

(lith: lithic, petri: petric, salin: saline, ptrcl: petrocalcic, ptrgp: petrogypsic, phrti: phreatic, gyp: gypsic)

**Table 11.28** Field properties of a representative phreatic Calcic Yermosol [Hypocalcic Calcisol] (No. 080) out of alluvial sand

| Depth (cm) | Horizon | Field properties |
|---|---|---|
| 0–0.5 | D | 7.5YR7/6, S, 24% Gravel and stones, single grain, very loose, contains carbonate, contains some carbonate concretions, fine layered, $EC_s = 3$ mS/cm |
| 0.5–15 | Ak | 7.5YR7/6, S, 4% Gravel and stones, gritty to platy, very weak, loose, shows remains of insects, rich in carbonate, contains calcareous concretions, $EC_s = 1$ mS/cm |
| 15–40 | Bwk | 10YR8/3, S, 19% Gravel and stones, polyeder to gritty, weak, dense, very rich in carbonate, rich in calcareous concretions, $EC_s = 4$ mS/cm |
| 40–85 | Bk | 10YR8/3, S, 19% Gravel and stones, polyeder to gritty, hard, dense, very rich in carbonate, rich in calcareous concretions, $EC_s = 5$ mS/cm |
| 85–150 | Ck | 10YR8/4, S, 27% Gravel and stones, polyeder to gritty, hard, dense, very rich in carbonate, rich in calcareous concretions, contains some roots, $EC_s = 4$ mS/cm |
| 150–180 | Ck | 10YR8/4, S, 12% Gravel and stones, polyeder to gritty, hard, dense, contains carbonate, rich in calcareous concretions, $EC_s = 6$ mS/cm |

**Table 11.29** Some chemical properties of a representative phreatic Calcic Yermosol [Hypocalcic Calcisol] from alluvial sand

| No. | Depth (cm) | Hor | $EC_d$ (mS/cm) | $S_t$ (‰) | N (ppm) | $CaCO_3$ (%) | $Fe_o$ (‰) | $Fe_t$ (‰) |
|---|---|---|---|---|---|---|---|---|
| 080.01 | 0–0.5 | D | 0.090 | 0.25 | 0 | 1.13 | 0.073 | 3.150 |
| 080.02 | 0.5–15 | Ak | 0.050 | 0.14 | 0 | 2.17 | 0.067 | 4.270 |
| 080.03 | 15–40 | Bwk | 0.190 | 5.38 | 0 | 11.33 | 0.016 | |
| 080.04 | 40–85 | Bk | 0.240 | 0.39 | 0 | 8.02 | 0.016 | 3.350 |
| 080.05 | 85–150 | Ck | 0.220 | 0.08 | 0 | 8.91 | 0.013 | 4.100 |
| 080.06 | 150–180 | Ck | 0.290 | 0.30 | 0 | 1.88 | 0.021 | 4.050 |

The field properties of a terrestrial Calcic Yermosol out of basalt at the Gilf Kebir are presented in Table 11.31. Results of chemical analysis in Table 11.32 show that the Bwk horizon is enriched with salt, gypsum, and carbonate. The $Fe_o$:$Fe_t$ ratio is 0.028 for the Aw horizon and 0.006 for the Bwk horizon. This indicates that the Aw horizon is more influenced by alteration than the Bwk horizon. Compared to the phreatic Calcic Yermosol, the terrestrial variant is enriched with nitrogen.

**Photo 11.16** Calcic Yermosol from basalt with stony, petric, and lithic phases

**Table 11.30** Number of terrestrial Calcic Yermosol variants and their "phases"

| n | lith | petri | stony | salin | ptrsl | ptrcl | ptrgp | phrti | gyp |
|---|---|---|---|---|---|---|---|---|---|
| 4 | | | | | | | | | |
| 2 | + | | + | | | | | | |
| 1 | | | + | + | | | | | |
| 7 | 2 | 0 | 3 | 1 | 0 | 0 | 0 | 0 | 0 |

(lith: lithic, petri: petric, salin: saline, ptrcl: petrocalcic, ptrgp: petrogypsic, phrti: phreatic, gyp: gypsic)

**Table 11.31** Field properties of a terrestrial Calcic Yermosol (No. 156) [Leptic Calcisol] out of basalt

| Depth (cm) | Horizon | Field properties |
|---|---|---|
| 0–1 | D | Basalt gravel and aeolian sand, mS, single grain, loose |
| 1–10 | Aw | 7.5YR5/6, lS, 30% Gravel and stones, platy to polyeder, very weak, upper limit spongy, ECs = 3 mS / cm |
| 10–40 | Bwk | SL, 70% Stones, gritty, very weak, rich in carbonate, large calcareous concretions, $EC_S$ = 5 mS/cm |
| > 40 | R | Basalt rock |

**Table 11.32** Some chemical properties of the fine earth of a terrestrial Calcic Yermosol [Leptic Calcisol] out of basalt

| No. | Depth (cm) | Hor | $EC_d$ (mS/cm) | $S_t$ (‰) | $N_t$ (ppm) | $CaCO_3$ (%) | $Fe_o$ (‰) | $Fe_t$ (‰) |
|---|---|---|---|---|---|---|---|---|
| 156.01 | 1–10 | Aw | 0.114 | 0.20 | 30 | 0.43 | 0.107 | 3.80 |
| 156.02 | 10–40 | Bwk | 0.216 | 0.25 | 60 | 4.30 | 0.282 | 43.75 |
| 156.03 | > 40 | R | | | | | | |

## 11.7 Chromic Vertisol (Vc)

The Chromic Vertisol is a clay-rich and self-mulching soil. In the dry state, it has cracks (Photo 11.17) and shows slickensides in the subsoil (Photo 11.18). Only Chromic Vertisols were identified in the field (Photo 11.19). This soil unit can be found in shallow depressions on playa sediments in the foreland of escarpments as well as on plains of clay-rich Mesozoic sediments. It is particularly common in Garb El Mauhub–Dakhla–Kharga–Baris area, as well as in the foreland of Abu Ballas escarpment, and at the foot of basalt cones. According to the 1998 WRB, the Chromic Vertisol of the investigated region is divided into three soil units: Salic, Eutric, and Gypsic Vertisol.

The Chromic Vertisol is often strongly enriched with salts, which indicates that the peloturbation characteristics or the vertic phase are relict and the "slickensides" was formed in the pluvial times. In addition, the slickensides usually occur in

**Photo 11.17** >1 cm wide crack in a Chromic Vertisol near the Baris oasis

**Photo 11.18** Slickenside on the surface of an aggregate from a Vertisol

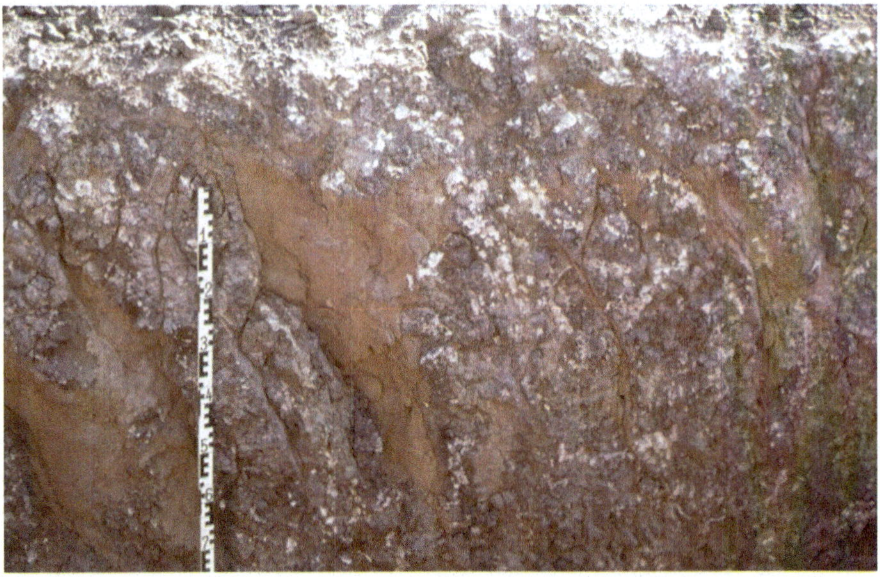

**Photo 11.19** Chromic Vertisol with cracks filled with aeolian sand

the upper soil (depth < 25 cm) and even at the soil surface. This shows that the investigated Vertisol is eroded to varying degrees by wind under the present climatic conditions (Photo 9.9). The deepest observed "slickensides" was up to 70 cm depth and hydro-cracks up to 160 cm depth east of the oasis Bulaq (Smettan 1987).

In addition to the saline phase, some Vertisols have a "gypsic phase" (Table 11.33).

Most of the Vertisols are deep, very poor in stones and rarely covered with stones or with aeolian sand. Table 11.33 shows that some Vertisols have no phases. In the current aridity Vertisol out of basaltic colluvium is extremely hard. They could not even be dug up with our excavator.

Table 11.34 shows the field characteristics of a typical representative of the Chromic Vertisols. The parent material of this soil is clay-rich colluvium from the Dakhla Formation, and it is located in a depression with elongated Yardangs, in the foreland of the Egyptian escarpment (Abu Tartur). The soil is covered with very loose fine gravel, and the soil surface is cracked by 3–5 cm wide cracks arranged in polygonal patterns.

The results of laboratory analysis from a representative Chromic Vertisol in Table 11.35 show that the soil is very rich in clay and has high electrical conductivity and high amount of sulfur as well as high amount of nitrogen. The electrical conductivity decreases with depth: this indicates a high atmogenic salt enrichment. With the soluble salts, nitrogen, probably as nitrate, is enriched in the soil. The low water conductivity and the high-water capacity of the soil prevented the salts from

**Table 11.33** Number of Chromic Vertisol variants and their "phases"

| n | lith | petri | stony | salin | ptrsl | ptrcl | ptrgp | phrti | gyp |
|---|---|---|---|---|---|---|---|---|---|
| 5 | | | | + | | | | | |
| 2 | | | | + | | | | | + |
| 1 | | | + | + | | | | | + |
| 1 | | | | | | | | | + |
| 1 | | | | | | | | | |
| 10 | 0 | 0 | 1 | 8 | 0 | 0 | 0 | 0 | 4 |

(lith: lithic, petri: petric, salin: saline, ptrcl: petrocalcic, ptrgp: petrogypsic, phrti: phreatic, gyp: gypsic)

**Table 11.34** Field properties of a Chromic Vertisol [Salic Vertisol] (No. 409) from playa sediment

| Depth (cm) | Horizon | Field properties |
|---|---|---|
| 0–5 | A | 10YR6/4, loamy clay, loose to moderately dense, platy, moderately hard, contains some gypsum and some carbonate, $EC_S = 63$ mS/cm |
| 5–26 | C | 2.5Y5/5, Loamy clay, very dense, roughly polyeder, very hard, has slickensides, contains some gypsum and some carbonate, $EC_S = 37$ mS/cm |
| 26–60 | C | 2.5Y6/5, Clay, very dense, prismatic, very hard, has slickensides, contains some gypsum and little carbonates, $EC_S = 39$ mS/cm |

**Table 11.35** Analysis data of the fine earth of a Chromic Vertisol [Salic Vertisol] out of playa sediment

| No. | Depth (cm) | Hor | $EC_d$ (mS/cm) | $S_t$ (‰) | $N_t$ (ppm) | $CaCO_3$ (%) | $Fe_o$ (‰) | Clay (%) |
|---|---|---|---|---|---|---|---|---|
| 409.01 | 0–5 | A | 11.290 | 11.21 | 1150 | 2.45 | 0.441 | 43 |
| 409.02 | 5–26 | C | 6.710 | 11.11 | 500 | 2.32 | 0.449 | 47 |
| 409.03 | 26–60 | C | 7.060 | | 600 | 2.48 | 0.418 | 51 |

leaching and caused their accumulation on the topsoil. The depth functions of the $Fe_o$ and carbonate contents show no tendency with depth, which indicates that the soil is hardly affected by alteration.

## 11.8 Other Soils

In addition to the above-mentioned soil units, the following soils occur on small areas:

- Gypsic Yermosol (Yy): is mainly found on "Desert Rose" and "Lower Clastic" formations. According to the WRB, the Gypsic Yermosol is named Gypsisol, forming a "major soil grouping" that includes several soil units.
- Calcaric Regosols (Re): are found on calcareous recent aeolian sands in the area of Tarfawi and in ergs of the eastern part of the investigated region.
- Eutric Fluvisol (Je): occur in some channels in the wadis of Gilf Kebir (Photo 11.20). Due to the rarity of rainfall, this soil is classified as Eutric Regosols.
- Calcaric Fluvisol (Jc): are found in some channels in the wadis of limestone escarpment (e.g., Abu Tartur escarpment) (Photo 11.21). Since precipitation or surface water occurs only in decades, this soil can be classified as Calcaric Regosol.
- Eutric Gleysol (Ge) and Calcaric Gleysol (Gc): occur in some hollows in the Kharga Oases.
- Takyric Yermosols (Yt) and Takyric Solonchaks (Zt): are located on playa sediments in the underruns of some wadis in the area of the Gilf Kebir and the Abu Tartur escarpment (Photo 11.22). They can be found after rain events; otherwise, the takyric structure is usually destroyed by wind (Photos 11.23 and 11.24).

Apart from the Takyric Solonchak, the equivalent soil units according to WRB of 1998 are shown in Table 11.10.

**Photo 11.20** Eutric Fluvisol or Eutric Regosol from fluvial sediment in a dry water channel in the middle reaches of Wadi Maftuh in Gilf Kebir

**Photo 11.21** Calcaric Fluvisol or Calcaric Regosol out of gravelly wadi sediment

**Photo 11.22** Takyric Yermosol from playa sediment in the lower reaches from a wadi south of the North Plateau

**Photo 11.23** Relatively freshly formed takyric structure on a playa sediment south of the northern plateau

**Photo 11.24** Remains of takyric structure from playa sediment of a final pan in the lower reaches of Wadi Wassa southeast of Gilf Kebir

## References

Alaily F (1993) Soil association and land suitability maps of the western desert, SW Egypt. In: Meissner B, Wycisk P (eds) Geopotential and ecology—analysis of a desert region. Catena Supplement, vol 26, pp 100–123

FAO-UNESCO (1974) Soil map of the world 1:5000000, vol I. Legend, Unesco Paris

ISSS, ISRIC, FAO (1998) World reference base for soil resources. FAO, World soul resources report, No. 84, Rome

Smettan U (1987) Typische Böden und Bodengesellschaften der Extremwüste Südwest-Ägyptens. Berliner geow Abh A 83:1–190

# Chapter 12
# Soil Landscapes

**Abstract** The region was divided into seven large geomorphological units and 28 soil associations. The geomorphological units represent different soil landscapes. Serir landscape with nine9 soil associations. The parent material of the soils consists mainly of alluvial sands, which are often covered by aeolian sand. Mesozoic rocks or Quarzitic stones appear sporadically as small outcrops. Groundwater occurs a few centimeters depth at Bir Tarfawi. Erg landscape with one soil association. The parent material of the soils of this landscape consists mainly of aeolian sand sheets and dunes. Between the dunes, there is sometimes alluvial sand. Deflation plains with five soil associations. The parent material of the soil consists mainly of various sandstones, silt stones, quarzitic-sandstones, gneiss, and granite. Hilly landscapes with six soil associations. These landscapes consist of inselbergs of Mesozoic sandstones and silt stones as well as hills of basalt and quartzite sandstones. Limestone plateau and its escarpment with two soil associations. This landscape is located north of Dakhla-Kharga and east of Kharga-Baris. It consists of limestone plateau, which is bordered by steep slopes. The slopes are cut by wadis filled with fluvial sand and gravel. The wadis end in deflated-depressions, leveled by playa-sediments and interrupted by inselbergs of Mesozoic sandstones and silt stones. The playa sediments are partly covered with aeolian sand and with various dunes. Sandstone plateau and its escarpment with four soil associations. The parent material of the soils on the plateau consists of Jurassic and Cretaceous sand as well as silt stones. Quaternary sandy and silty sediments are found in small shallow depressions and dry water channels. Quarzitic sandstones and basalt mounds are scattered on the plateau. Oases landscape with one soil association. This landscape consists of playa, fluvial, and aeolian sediments with sandy to clayey soils.

**Keywords** Soil association · Serir · Bir Tarfawi · Erg · Deflation plains · Limestone plateau · Sandstone plateau · Oases landscape · Alluvial sand · Aeolian sand · Quarzitic stones · Mesozoic rocks · Sandstones · Silt stones · Gneiss · Granite · Basalt



F. Alaily, *Ecosystems of Extremes*, Advances in Geographical and Environmental Sciences, https://doi.org/10.1007/978-3-031-76910-8_12

Based on our field results and with the help of the geological maps of SW Egypt with a scale of 1: 500,000 (Klitzsch and List 1980) and through visual and digital interpretation of LANDSAT multispectral images, the region of southwest Egypt was divided into seven large geomorphological units (Alaily 1993) and 28 soil associations. The geomorphological units represent different soil landscapes. The distribution pattern of the geomorphological units and soil associations are shown on a map with a scale of 1:1,000,000 (Alaily and Blume 1986). In addition to soil units, the soil parent material and terrain morphology with combinations of the "soil phases" from the dominant soil of each soil association (given in small letters) are indicated on the map. This allows the map to be interpreted in soil genetic and ecological terms. The soil association map of southwest Egypt is shown in Fig. 12.1 with a simplified legend.

## 12.1 Serirs

Serirs represent the largest landscape unit in the investigated region. They are slightly undulating and are largely represented by the "Selima Sand Sheet" and the "Atmur El Kabeish". The parent material of the soils consists mainly of alluvial sands, which are often covered by aeolian sand. Mesozoic rocks or Quarzitic stones appear sporadically as small outcrops. The groundwater depth decreases from Gilf Kebir in direction of Bir Tarfawi and Bir Safsaf. Groundwater occurs a few centimeters depth at Bir Tarfawi and a few meters depth at Bir Safsaf. Tamarisk bushes (*Tamarix spp.*) and date palms (*Phoenix dactylifera*) characterize the landscape at Bir Tarfawi. Reeds (*Phragmites australis*), date palms, and dome palms (*hyphaene thebaica*) are found at Bir Safsaf. At Bir Sahara (28 km southwest of Bir Tarfawi) the groundwater occurs at a depth of 30 m and the area is free of vegetation. The relationship between the soil unit, parent material, groundwater level, and vegetation on the Selima Serir can be seen in Fig. 12.2.

Haplic Yermosol (Yh), Calcic Yermosol (Yk), and Orthic Solonchak (Zo) occur on alluvial sand. In the order mentioned, the groundwater depth decreases. On the Tamarisk hills are Orthic Solonchak to be found and in the Birs is Gleyic Solonchak. In the surroundings of the Birs, there are Orthic Solonchak and Calcic Yermosol, which are "petric" and partly "stony" because of their high content on calcareous concretions. The Calcic Yermosol is "saline" and the Orthic Solonchak is partly "petrosaline" and "phreatic".

Cambic Arenosol and Eutric Regosol are found on aeolian sand of different ages (Alaily and Blume 1983 and Blume et al. 1984). Calcaric Regosol consisting of aeolian sand (50–100 cm thick) over Calcic Yermosol is found in the vicinity of the Birs. The Cambic Arenosol, Eutric Regosol, and Calcaric Regosol are deep and have no phases. On the Mesozoic sand and siltstones that occur sporadically in the Serir areas, Lithosol is found, that are partly "saline". Apart from the Lithosol, all soils in this landscape are sandy.

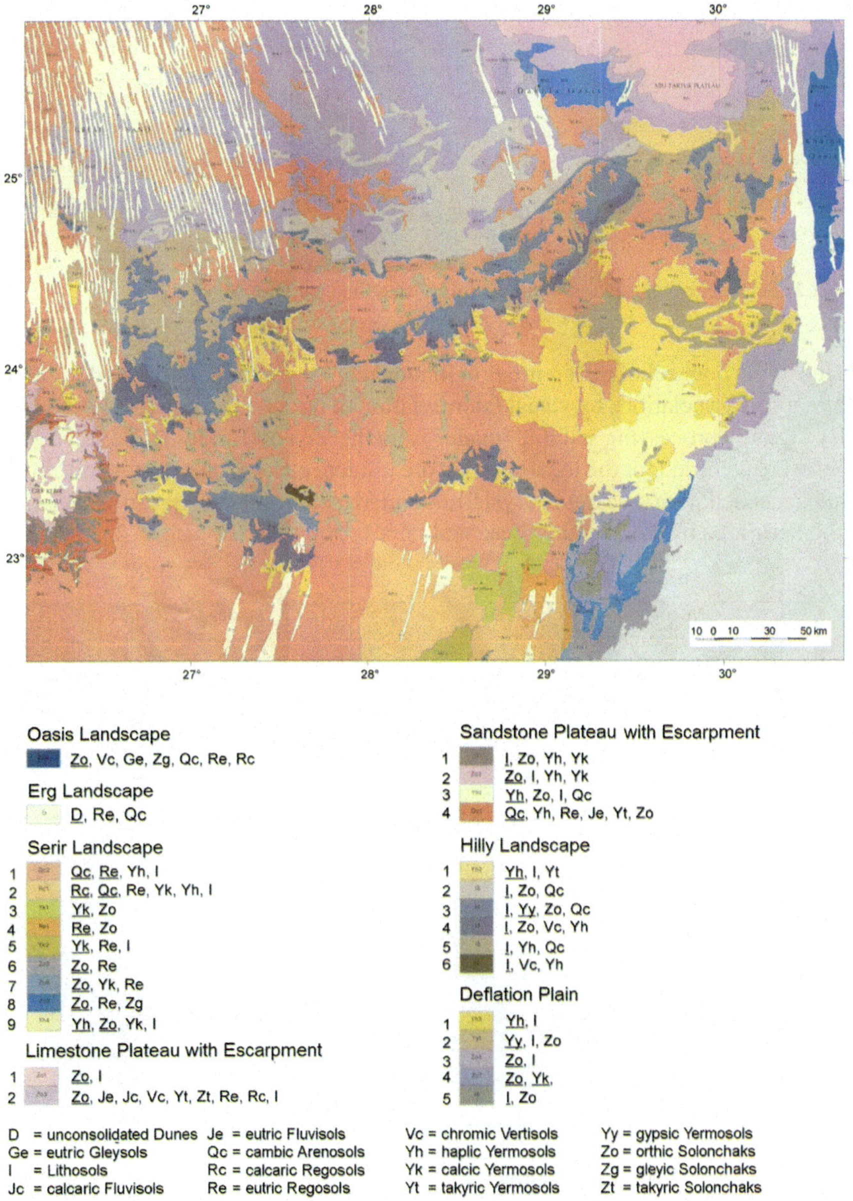


**Fig. 12.1** Soil association map of Southwest Egypt with simplified legend

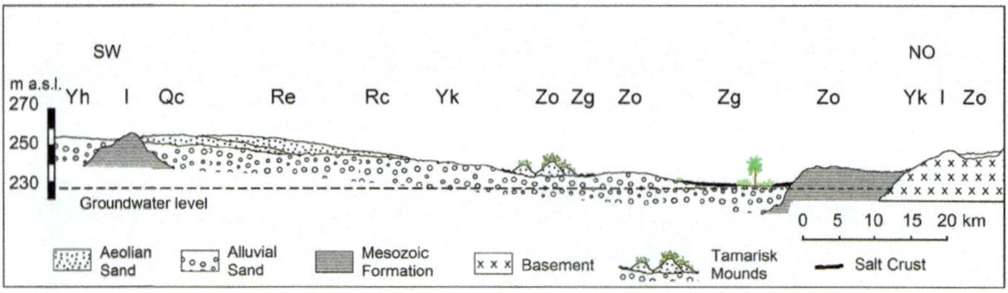


**Fig. 12.2** Schematic presentation of the Selima Serir soil landscape

Due to the poor relief differentiation in the area, it was not possible to delimit the different soil associations by visual interpretation of the satellite images (LANDSAT I). By digitally processing satellite images from the area by Tarfawi; nine soil associations were recorded for the locations that are influenced by groundwater and two for the adjacent deflation area out of crystalline and Mesozoic rocks. All soil associations on the Serir area by Tarfawi are presented on a 1: 500,000 map (Fig. 12.3).

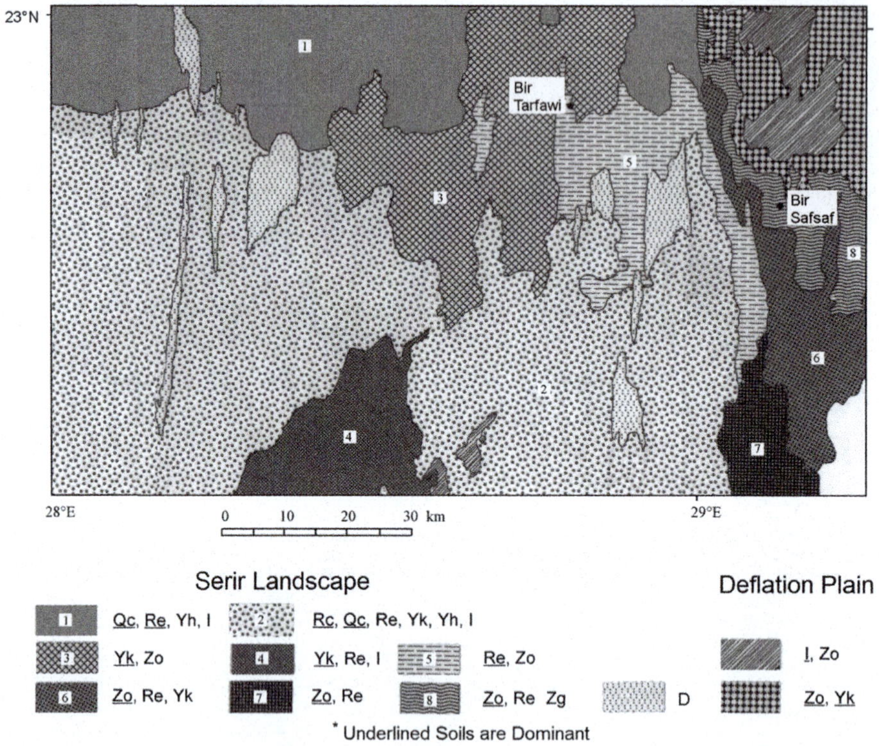


**Fig. 12.3** Soil association map for the Tarfawi area with a scale of 1: 500,000 (Alaily and Pohlmann 1995)

On the soil map in Figs. 12.1 and 12.3, the following soil associations on the Serir landscape (soils in bold are dominant) are presented:

**Cambic Arenosol, Eutric Regosol** out of aeolian sand; Haplic Yermosol out of alluvial sand in shallow depressions and between sand hills; Lithosol on outcrop-rock. This soil association covers the largest area on the Serir;
**Calcaric Regosol, Cambic Arenosol,** and Eutric Regosol from aeolian sand; Calcic Yermosol and Haplic Yermosol from fluvial sand between sand hills; Lithosol on outcrop-rock.
**Calcic Yermosol** from fluvial sediments on plains; Orthic Solonchak in shallow sinks,
**Eutric Regosol** from aeolian sand, Orthic Solonchak from fluvial sediments between sand hills,
**Calcic Yermosol** from fluvial sediments on plains; Eutric Regosol on sand hills, Lithosol on outcrop-rock,
Carbonate concretion-rich **Orthic Solonchak** out of fluvial sediments on plains and Calcic Yermosol on elevated sites as well as Eutric Regosol from aeolian sand hills.
Carbonate-rich **Orthic Solonchak** from fluvial sediments on plains, Eutric Regosol from aeolian sand hills (dunes).
**Orthic Solonchak** (often petrosaline) out of fluvial sand on plains, Eutric Regosol out of aeolian sand; Gleyic Solonchak at the waterhole (Bir),
**Haplic Yermosol and Orthic Solonchak** out of fluvial sediments on all relief positions, Calcic Yermosol in some shallow sinks, Lithosol on outcrop-rocks.

## 12.2 Ergs

The parent material of the soils of this landscape unit consists mainly of aeolian sand sheets and dunes. Between the dunes, there is sometimes alluvial sand. This landscape consists of only one soil association dominated by **dunes**. Eutric Regosol and Cambic Arenosol are found between the dunes on the older aeolian sand and alluvial sand (Fig. 12.4). Occasionally one finds Haplic Yermosol on playa sediments and rarely Lithosol on Mesozoic rocks. The soils in this landscape have no phases apart from a “saline” phase, which occurs sporadically.

## 12.3 Deflation Plains

This landscape has a flat to slightly undulated surface. Occasionally there are traces of shallow dry water channels as well as small dents, thresholds, and steps. The parent material of the soil consists mainly of various sandstones, silt stones, quarzitic sandstones, gneiss, and granite.

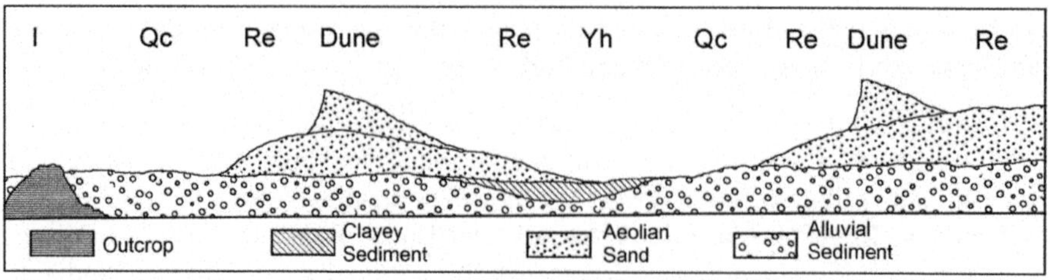


**Fig. 12.4** Schematic presentation of an erg soil association

On the soil association map of SW Egypt (Fig. 12.1), five soil associations are presented on the deflation plains:

- On sandstones of Lower Clastic formations **Haplic Yermosol** with lithic, stony, and petric phases occur on large areas. In addition, Lithosol can be found on heavily eroded surfaces.
- On Desert Rose sandstones **Gypsic Yermosol** with lithic, stony, petric, and saline phases are found as well as Orthic Solonchak, which are also lithic, stony, and petric. Lithosol occurs on heavily eroded surfaces.
- On Variegated Shale and on the Dakhla Formation **Orthic Solonchak** with lithic, petric, stony, and gypsic phases are common. Lithosol occurs on heavily eroded surfaces.
- On the Lower Clastic Formation and Granite, in southeast of the investigated region, **Orthic Solonchak** and **Calcic Yermosol** with predominantly lithic, stony, and petric phases are found. Only soils from Lower Clastic can have a gypsic phase.
- On granite, **Lithosol** are dominant, they are associated with Orthic Solonchak in small hollows. The latter are occasionally lithic and stony.

## 12.4 Hilly Landscapes

These landscapes consist of inselbergs of Mesozoic sandstones and silt stones as well as hills of basalt and quartzite sandstones. All hills are covered with stones, iron concretions, or remnants of iron crusts. The stones are heavily blackened by desert varnish and therefore have a similar color to iron-concretions and iron-crust remains. The hills have an average height of 100–300 m and are scattered across the landscape. Between the hills, there are rubble, aeolian, fluvial, and playa sediments.

Because of the high diversity of the parent material and relief, the hilly landscapes have numerous soil units with different phases. Only the most common soil units and their phases are presented in this chapter.

Lithosol is the most common soil in almost all hilly landscapes. They are found on all steep slopes as well as on humps and in deflation aisles between the hills. Cambic Arenosol from aeolian sand is also widespread in all hilly landscapes. They are found in small sites mainly on the lee side of the hills. In addition to Lithosol and Cambic Arenosol, the following soil units occur in the soil associations of the various geological formations. Apart from the first soil association (see below), **Lithosol** is the dominant soil in the other soil associations.

(1) Between Dakhla and Kharga, the hilly landscapes of Desert Rose have **Haplic Yermosol** and Takyric Yermosol from fluvial sediments. Both soils have no phases. In addition, Haplic Yermosol from rubble which is stony, petric, and lithic is found.
(2) Hilly landscapes from Plant Bed's formations have Orthic Solonchak from rubble as well as on pediments and occasionally Cambic Arenosols from fluvial sediments. The Orthic Solonchak on pediment are gypsic and stony. Those on rubble are gypsic and often stony, petric, and sometimes lithic.
(3) In addition to the soil associations described above under (1), hilly landscapes from the Desert Rose Formation have Gypsic Yermosol on rubble and fluvial sediments and Orthic Solonchak on plateaus. The Orthic Solonchak are lithic, stony, and petric. The Gypsic Yermosol is stony and saline.
(4) Hilly landscape from Lingula Shale has lithic, petric, stony, and occasionally gypsic Orthic Solonchak out of rubble. Chromic Vertisol and Haplic Yermosol are found on playa sediments. Haplic Yermosol and Cambic Arenosol occur on fluvial sand.
(5) Hilly landscapes from the Lower Clastic Formations have Haplic Yermosol on debris and on fluvial sediments and Cambic Arenosol on fluvial sediment. The Haplic Yermosol is stony, petric, lithic, and occasionally gypsic. The Cambic Arenosol is occasionally lithic.
(6) Hilly landscapes from basalt and quartzite hills are found in the south and southwest of the investigated region. They have Chromic Vertisols on playa sediments and Haplic Yermosol from fluvial deposits. Both soils are sometimes saline.

## 12.5 Limestone Plateau and Its Escarpment

This landscape is located north of Dakhla-Kharga and east of Kharga-Baris. It consists of a high limestone plateau 500 m a. s. l., which is bordered by steep slopes to the south (north of Dakhla-Kharga) and west (east of Kharga-Baris). The slopes are cut by wadis filled with fluvial sand and gravel. The wadis end in deflated depressions, leveled by playa sediments and interrupted by inselbergs of Mesozoic sandstones and silt stones. The playa sediments are partly covered with aeolian sand and with various dunes.

The soils of this landscape are grouped into two soil associations on the map of southwestern Egypt. A schematic section of a wadi in the Limestone plateau and its vorland are presented in Figs. 12.5 and 12.6.

The illustrations show that: on the plateau petric, stony, often gypsic and lithic **Orthic Solonchak** is found. The stones on the surface are colored grey by desert varnish. Orthic Solonchaks, which are petric, stony, and gypsic, can be found at the foot of the hillside debris and on wadi terraces. Calcaric Regosol (or Calcaric Fluvisol) and occasionally Eutric Regosol (or Eutric Fluvisol) occur in the wadi bed. Both soils are lithic, saline, often stony and petric.

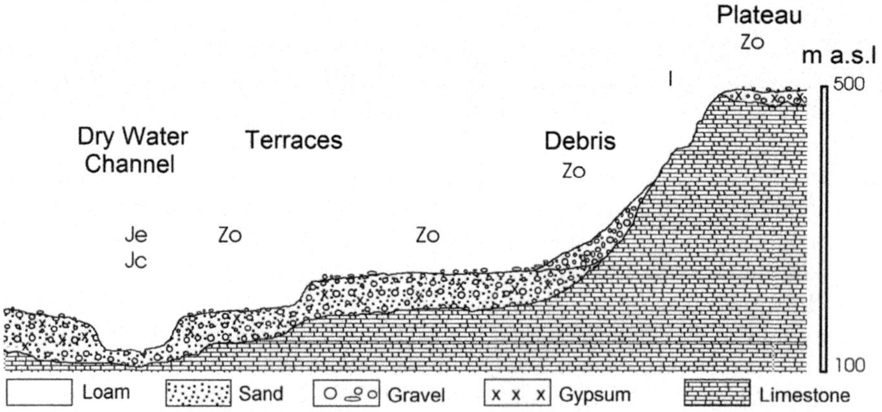


**Fig. 12.5** Schematic presentation of the soil association on the limestone plateau and the wadis

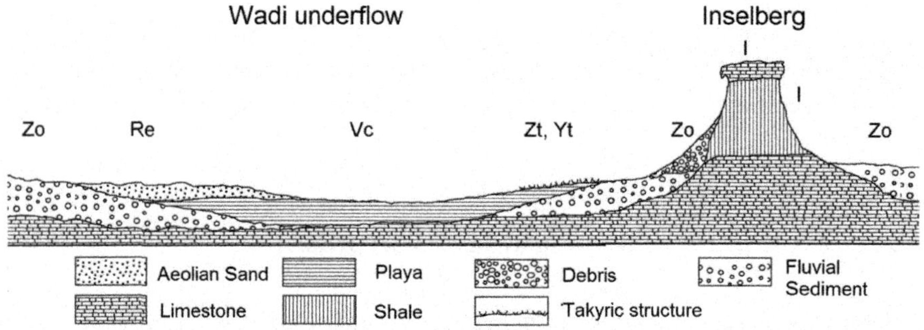


**Fig. 12.6** Schematic presentation of the soil association in a wadi lower-run (underflow) of the limestone plateau

In the end pans of the wadi underflows we find Chromic Vertisol with "high salinity" on clay-rich playa sediments. On playa sediments with less clay content occur Takyric Solonchak and Takyric Yermosol with a saline phase. In the vicinity of the playas, stony, gypsic, occasionally petric, and lithic Orthic Solonchaks out of Mesozoic siltstones and sandstones occur. Eutric and Calcaric Regosol are found on aeolian sand. This soil unit has no phases. Lithosol occur mainly on the steep slopes as well as on outcrop rocks in the heavily eroded part of the Wadis.

## 12.6 Sandstone Plateau and Its Escarpment

This landscape occurs in southwest of the investigated region in the Gilf Kebir. The plateau of the Gilf Kebir is roughly circular, with a diameter of about 100 km. It has a slightly agitated relief and is surrounded by numerous wadis with steep slopes. The average height of the plateau is 900–1050 a.s.l. The parent material of the soils on the plateau consists of Jurassic and Cretaceous sand as well as silt stones. Quaternary sandy and silty sediments are found in small shallow depressions and dry water channels. Quarzitic sandstones and basalt mounds are scattered on the plateau.

The wadis are usually filled with sandy sediments. Occasionally, terraces of different pluvial times can be found. The slopes are covered with massive debris and often with aeolian sand. In some narrow wadis, the parent material of the soil consists entirely of rubble. In west–east-facing wadis there are barrier dunes. Playas sediments can often be found behind such dunes, as well as in wadi locations that suddenly become wider.

Figure 12.7 presents schematically the soil landscape on the plateau and in wadis. This shows that on the plateau Orthic Solonchak, Haplic Yermosol, and Lithosol are found. The Orthic Solonchak is mainly found on flat surfaces and is lithic, stony, and petric. The Haplic Yermosols can be found in small depressions and in dry water channels. They are "saline", often "petric", "lithic" and occasionally "stony". On basalt hills, one finds Calcic Yermosol, which is lithic, petric, and stony. Lithosol can be found on steep slopes of basalt or quarzitic sandstone. Cambic Arenosol occupy small patches, which are covered with aeolian sand. The stones of the soil cover on the plateau are heavily blackened by desert varnish.

At the foot of the slope in the wadis, Cambic Arenosol and Haplic Yermosol from debris are found. Cambic Arenosol is also found on terraces and aeolian sand deposits. In the wadi bed and in the upper wadi areas, Eutric Regosol and Eutric Fluvisol are found. On playa sediments, Haplic Yermosol, Takyric Yermosol, and, rarely, Orthotic Solonchak are found. The soils in the upper reaches of the wadis are mostly stony and petric, those found on the terraces are occasionally stony and rarely saline, and those in the wadi bed are occasionally stony and lithic. The Haplic and Takyric Yermosol from playa sediments are mostly saline.

On the soil association map of SW Egypt with a scale of 1: 1,000,000, this landscape consists of four soil associations (Fig. 12.1), which were delimited by means of digital and visual interpretation of LANDSAT images (Alaily et al. 1987, Alaily

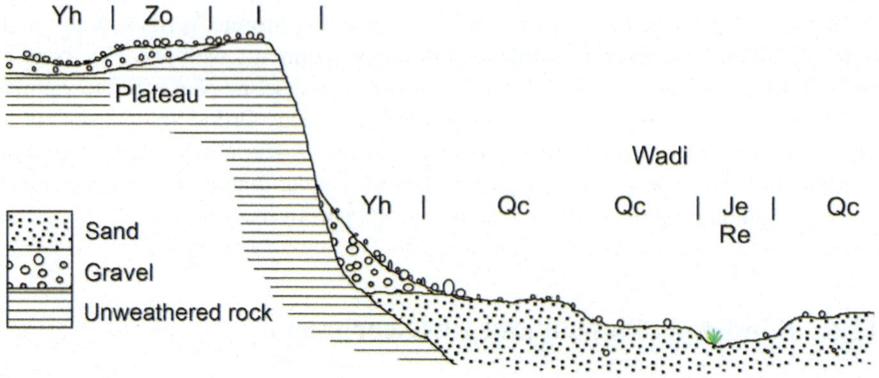


**Fig. 12.7** Schematic presentation of the soil association on the plateau and in the wadis of the Gilf Kebir

and Pohlmann 1983). Because of the small scale of the map, only one soil association could be drawn for the wadis. In the following list, the first three soil associations are found on the plateau and the fourth in the wadis:

- **Lithosol** is found on steep slopes. Orthic Solonchak and Haplic Yermosol from Mesozoic rocks are found on plains or in small hollows and Calcic Yermosol on basalt humps.
- On Mesozoic rocks: **Orthic Solonchak** on plains, Lithosol on small humps, Haplic Yermosol in shallow depression, and Calcic Yermosol on basalt humps.
- **Haplic Yermosol** and Orthic Solonchak from kolluvium are found in large shallow depressions, Lithosol on humps of Mesozoic rocks, Cambic Arenosol on plains of fluvial sand.
- **Cambic Arenosol** and Haplic Yermosol from terrace sand and debris, Eutric Regosol, and Eutric Fluvisol from fluvial sand in wadi channels (wadi bed) are found. Cambic Arenosol and Eutric Regosol from aeolian sand. Takyric Yermosol and Orthic Solonchak from playa sediments in the lower reaches of the wadis and behind some barrier dunes.

## 12.7 Oases

This landscape is represented on the map with only one soil association. It consists of playa, fluvial, and aeolian sediments with sandy to clayey soils (Fig. 12.8). In this soil association, **Orthic Solonchak** and **Chromic Vertisol** are very common on playa and fluvial sediments. Eutric Gleysol and Gleyic Solonchak occur in hollows with groundwater connection. Cambic Arenosol, Eutric Regosol, and Calcaric Regosol from aeolian sand are found on small isolated locations. All the soils in this association are deep, poor in stones, and have only “saline phase”.

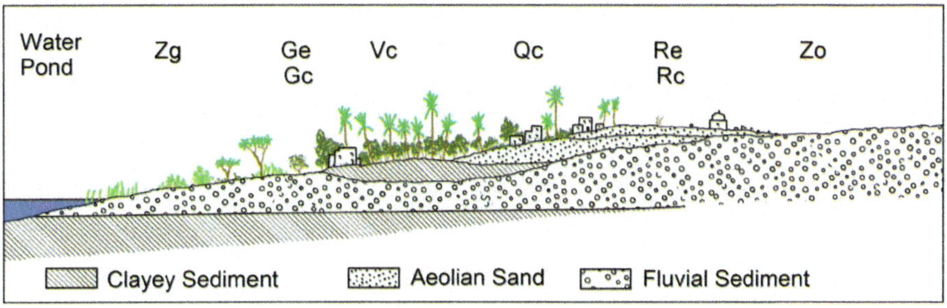


**Fig. 12.8** Schematic presentation of an oasis soil association

# References

Alaily F (1993) Soil association and land suitability maps of the western desert, SW Egypt. In: Meissner B, Wycisk P (eds) Geopotential and ecology—analysis of a desert region. Catena Supplement, vol 26, pp 100–123

Alaily F, Blume H-P (1983) Bodengesellschaft der Vollwüste, Dakhlabecken—Ägypten. Mitteilgn Dtsch Bodenkundl Gesellsch 38:443–449 (Göttingen)

Alaily F, Blume H-P (1986) Ecological maps of the western desert of Egypt soil association map. TFH-Berlin

Alaily F, Pohlmann J (1983) Eine flächenhafte Erfassung von Bodengesellschaften arider Gebiete mit Hilfe von LANDSAT-Multispektralaufnahmen. Berliner geowiss Abh (A) 47:49–61

Alaily F, Pohlmann J (1995) Soil mapping of Bir Tarfawi region (SW-Egypt) based on digital classification of landsat-MSS-data. Adv GeoEcology 28:89–107

Alaily F, Bornkamm R, Blume H-P, Kehl H, Zielinski H (1987) Ecological investigations in the Gilf Kebir. Phytocoenologia 15(1):1–20 (Stuttgart-Braunschweig)

Blume H-P, Alaily F, Smettan U, Zielinski H (1984) Soil types and associations of southwest Egypt. Berliner geowiss Abh (A) 50:293–302

FAO-UNESCO (1974) Soil map of the world (1:5000000). vol I, Legend, UNESCO Paris

Klitzsch E, List FK (1980) Egypt-geological interpretation map, 1:500000, (preliminary ed). TFH Berlin

# Chapter 13
# Ecological Systems

**Abstract** The investigated region is divided into autochthonous and allochthonous ecosystems. Most of the area has an allochthonous ecosystem. In this area, the existence of plants is not possible, but we still find consumers. The existence of such animals depends on food imports. The import is done by wind, migratory birds, and caravans. Depending on the amount and frequency of precipitation as well as groundwater level and land use, the autochthonous ecosystems is divided into: (1) permanent natural oases, which are (a) inhabited by humans and represented by the traditional oasis complexes Dakhla, Kharga, and Baris, (b) uninhabited by humans such as Bir Tarfawi, Bir Safsaf, and (2) permanent artificial oases that were created as part of land reclamation projects (East Oweinat Project: e.g., on the Selima Serir southwest of Bir Tarfawi), and (3) episodic oases (e.g., in depressions and wadis of the Gilf Kebir and the north and east plateaus). The duration (one season or several years) of these oases depends on the amount of water stored in the soil.

**Keywords** Allochthonous ecosystem · Natural oases · Uninhabited oases · Artificial oases · Episodic oases · Dakhla · Kharga · Baris · Bir Tarfawi · Bir Safsaf · East Oweinat · Gilf Kebir

An ecosystem is defined as the structure of effects of living beings and their inorganic environment that is open and, to a certain extent, capable of self-regulation (Bick 1989). Ecosystems consist of two main components: the ecocenoses (plants and animals or producers and consumers or destructors) and the ecotopes (soils, waters and airspace or pedo-, limno-, and atmotopes). The composition of the ecocenoses is determined by the characteristics of the ecotope. In the ecocenoses, plants are called producers because they can build organic components from inorganic substances through photosynthesis. Consumers are animals whose existence is directly (herbivores) or indirectly (carnivores) dependent on plants. The opposite pole to the producer is the destructors. Destructors include all organisms that can break down organic substances into inorganic components. Soil microorganisms in particular are



F. Alaily, *Ecosystems of Extremes*, Advances in Geographical and Environmental Sciences, https://doi.org/10.1007/978-3-031-76910-8_13

responsible for the breakdown and mineralization of organic matter. In contrast to consumers that can migrate to the water sources, plants and soil microorganisms have limited mobility. Their existence depends on the water supply in situ.

Because of the extremely dry climate in the studied region, producers are a great rarity. Locations with plants or soils with plant roots or root ducts that were observed in the period 1981–1986, are shown in Fig. 13.1. The figure shows that until 1986 green plants were mainly found permanently in the oases of Dakhla, Kharga, and Baris as well as in the Birs (uninhabited small oases). At present time, arable farming is carried out on a large scale on the Selima Serir, southwest of Bir Tarfawi. In all of these locations, the existence of plants depends on groundwater. They represent permanent ecosystems and are divided in: populated and unpopulated as well as natural and artificial ecosystems. In most of the investigated region, however, there is insufficient rainfall for producers to exist. Some plants can only grow in locations where the relief allows surface water from rain to accumulate (in depressions or wadis). The lifespan of the plants in these locations is only a few years. After that, the ecosystem is dormant until the next episodic rain event. Such an ecosystem is an episodic ecosystem (Kehl and Bornkamm 1993). In most of the area where the existence of plants is not possible, we still find consumers (animals such as the fennec, snakes, and lizards). The existence of such animals depends on food imports. The import is done by wind, migratory birds, and caravans. This ecosystem is an allochthonous ecosystem. In this way, the investigated region was divided into autochthonous and allochthonous ecosystems.

Depending on the amount and frequency of precipitation as well as groundwater level and land use, the autochthonous ecosystems is divided into:

(1) permanent natural oases, which are
    (a) inhabited by humans and represented by the traditional oasis complexes Dakhla, Kharga, and Baris.
    (b) uninhabited by humans such as Bir Tarfawi, Bir Safsaf, etc., and
(2) permanent artificial oases that were created as part of land reclamation projects (e.g., on the Selima Serir southwest of Bir Tarfawi), and
(3) episodic oases (e.g., in depressions and wadis of the Gilf Kebir and the north and east plateaus).

A comprehensive description of the ecosystems mentioned can be found in Kehl and Bornkamm (1993). In the following, all plants and animals are mentioned that were observed by the author in the above-mentioned ecosystems.

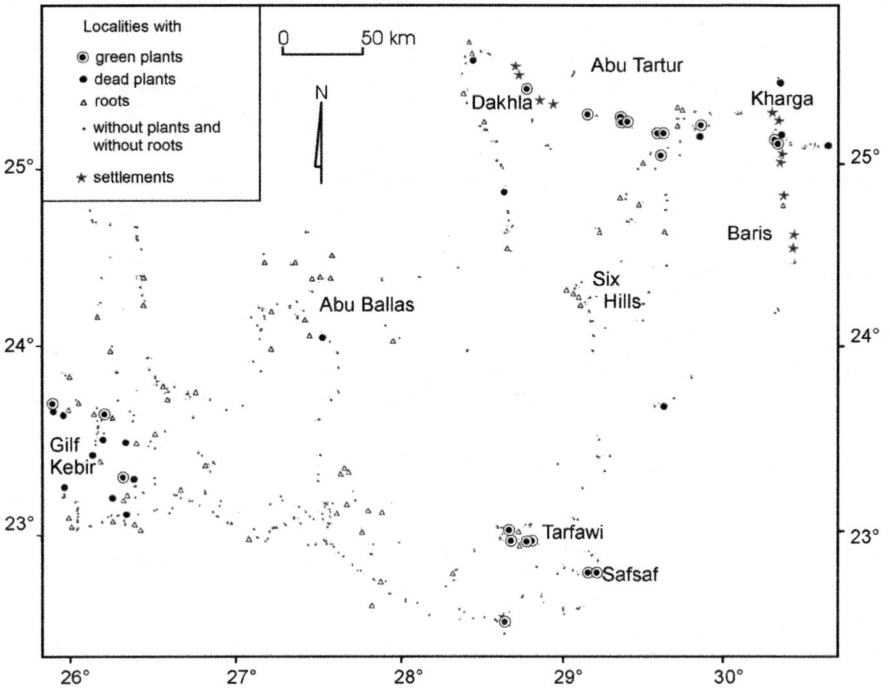


**Fig. 13.1** Localities of investigated soils with green plants, dead plants. and plant roots for the period 1981–1986

## 13.1 Permanent Autochthonous Ecosystem

### *13.1.1 Traditional Oases*

This ecosystem is represented by the oases complexes Dakhla, Kharga, and Baris. It is an anthropogenic ecosystem that is characterized by agricultural use and whose existence dependent on artificial irrigation. The Kharga–Baris oasis complex is 3693 km$^2$ in size. It is 38–80 m a.s.l. and has been settled since the Paleolithic. At the time of the pharaohs, the Kharga oasis was called Kenem or Hibis. According to the 1996 census, the city of El Kharga had 49,461 inhabitants. Baris is the second largest settlement in the Kharga Depression. It is located about 86 km south from El Kharga, and it has around 3900 inhabitants (census 2006). The Dakhla Oases lies in a depression 90–140 m a.s.l and 120 km west from El Kharga. It covers an area of 2000 km$^2$, in which around 75,000 people live (2003 estimate).

Cultivated plants (orchards and vegetable gardens as well as arable fields) and domestic animals dominate the landscape in the oases. The arable land is located in the Kharga Depression mainly around of city of El Kharga and the villages Maharik, Ginah, Bulaq, Baris, and 13 Ezbas (8 by El Kharga and 5 by Baris). In the Dakhla

Depression, arable fields are spread around the city of Mut and El Qasr as well as around 14 villages such as Balat, El Rashda, El Mushiya, and Ezbet El Sheikh Mawhub as well as 13 Ezbas. The total cultivated area is 12,915 ha (Euroconsult 1983). The fields and orchards are mainly irrigated by basins or furrow irrigation. In the orchards, the multi-story-cultivation method is used. Smaller plants are grown under the fruit trees (e.g., tomatoes, onions, beans, etc.). This way of cultivation is known as "agroforestry". Date palms (Phoenix dactylifera), figs, olives, oranges, grapes, bananas, and cereals (especially wheat and wet rice; Photo 13.1), various legumes (Vicia faba, Medicago sativa, etc.), cotton, and various types of vegetables are grown. Casuarina, eucalyptus, and tamarisk bushes are planted as windbreak trees. Dom palms (*Hyphaene thebaica*) can also be found in the Kharga–Baris oasis complex (Photo 13.2). This palm is missing in Dakhla.

In the vicinity of the oases, the soils are heavily enriched with salt. Such areas represent sebchas, which are caused by the recent formed (as a result of agricultural use) secondary groundwater level. Drainage water and the seepage water from the fields are often enriched in depressions around the oases in evaporation lakes or ponds (Photo 13.3). Salt-resistant plants such as *Tamarix spp., Juncus rigidur, Phragmites australis, Alhagi mannifera,* and *halophytes* can be found on the banks of such lakes and on the sebchas (Kehl and Bornkamm 1993). Due to the high evaporation, most evaporative ponds dry out in summer and a hard salt crust remains on the surface. In the Dakhla Oases, such lakes cover larger areas than in Kharga.

Consumers in this ecosystem include all domestic animals (camels, goats, sheep, cows, etc.). In 1983, there was a mice plague in Egypt as well as in the oases. Wild animals which have been observed by the author are: white herons, field foxes, and a one-meter-long varan as well as various insects (mosquito, various species of ants, termites, stick insects, hornets, grasshoppers, ant lions (Photos 13.4, 13.5 and 13.6).

### *13.1.2 Birs*

Birs are found in southeast of the investigated region. In most of the birs, the groundwater lies at a depth of a few centimeters to a few decimeters. At Bir Abu El Hussein, the groundwater is at a depth of 12 m, at Bir Sahara and Bir Misaha at about 30 and 18 m, respectively. In the bir areas (with the exception of Bir Sahara, Bir Misaha, and Bir Abu El Hussein), there is permanent vegetation connected to the groundwater. Due to bad soil conditions, the birs are not inhabited by people. The soil surroundings the birs is heavily enriched with salts and partly covered by an extremely hard and thick (25 cm) salt crust. The typical vegetation structure at the birs shows palm trees (*Phoenix dactylifera and Hyphaene thebaica* ) at the lowest points, followed by grasses or small shrubs (e.g., *Alhagi mannifera, Phragmites australis, Sporobolus spicatur, and Stipagrostis vulnerans*) (Darius 1989 and Schneider (1990) (Photos 13.7 and 13.8). As the distance to the groundwater increases, tamarisk mounds appear (Photo 13.9) and further away from the bir Acacia mounds occur (*Acacia ehrenbergiana*) (Photo 13.10) and *Ziziphus spina-christi* (Photo 13.11) (Kehl 1987). In Bir Tarfawi and Nakhla, the dom palm (*Hyphaene thebaica*) is missing. At Bir Nakhla, Schneider (1990) found *Imperata cylindrica*. A total of 16 plant species have been observed in the uninhabited oases (Kehl and Bornkamm 1993).

The groundwater depth under Acacia and Ziziphus hills as well as at some tamarisk hills is often more than 20 m. According to Back (1978), tamarisk roots (also called ethel trees) can reach a depth of thirty meters. If these plants were established in a humid climatic phase or in a phase with a high groundwater level, then these plants must be over 6000 years old and since then the roots have been growing behind the sinking groundwater.

In locations where plants have groundwater connection, the extremely low microbial degradation processes on the soil surface cause the accumulation of dead plant-leaves and branches, which ultimately form biogenic hills (Photos 13.12 and 13.13). Such mounds consist exclusively of plant remains and can reach a height of approx. 15 m (near Bir Kiseiba). The preservation of tamarisk remains is also favored by the high salt content of the litter.

From the density of insect traces and other animal tracks around birs, make such unique oases-ecotopes interesting for farther investigations. Not only do birs serve as a habitat for numerous animals, they are also important resting places for migratory birds and caravans. The only animals observed by the author were the fennec (Photo 13.14), the jerboa (*Jaculus jaculus*), and some raptors as well as a grasshopper and a 4-mm-long silver-pink-ant.

Numerous birs are currently threatened by groundwater-dependent agricultural projects. In the long term, lowering of groundwater and drying out of birs can be expected through such projects. In addition, the birs could be damaged or destroyed by:

1. leveling work for road construction,
2. collecting firewood (energy source for small farmers),
3. pets (such as dogs, cats, and goats),
4. burning the dry organic mounds to eliminate scorpions and snakes.

The latter happened in 1982: Land surveyors set Bir Tarfawi on fire to avoid possible danger from the animals mentioned (Photo 13.15).

### 13.1.3 Artificial Oases

These are oases that were created by "high technology" as part of the "Shark El Oweinat" project. In this project, an area of 4924 km$^2$ on the Selima Serir Plain is to be used for agriculture. First in 1980, an agricultural experimental farm was established on the Serir north of the Egyptian-Sudanese border. The aim of the experimental farm was to test various crops and irrigation systems. The necessary water was obtained through deep wells. The groundwater level was at a depth of 25 m. Wheat, onions, strawberries, peppers, corn, grapes, apricots, oranges, etc., were planted. Casuarina trees were used as wind protection. Drip irrigation, sprinkler, and basin irrigation as well as a combination of drip and basin irrigation were used. It was noticeable that hardly any weeds grew on the fields. The experimental farm was also a food source for autochthonous animals and migratory birds. The latter often stopped and stayed a few days in the experimental farm. In 1983, we observed that the jerboa population in the area had increased significantly compared to the year before.

Satellite images from 1999, 2001, and 2017 show that arable land on the Selima Serir is gradually increasing. Images taken by LANDSAT 8 on February 26, 2017, show a large number of circular fields, mainly west of the experimental farm (Fig. 13.2). These are irrigated by rotating sprinklers and with the fossil groundwater. Each of these circles has a diameter of about 1 km. The groundwater in the area is between 20 and 60 m under ground surface (EL Sayed et al. 2004), whereas the depth increases from east to west. How long this project can last depend on the groundwater availability and its consumption rate. With today's technical knowledge and economic conditions, Ebraheem et al. (2003) estimated the noneconomic groundwater extraction depth for the project area with > 100 m, and with a water consumption of $1200 \times 10^6$ $m^3$ $year^{-1}$, (as planned) this depth will be exceeded in the year 2100.

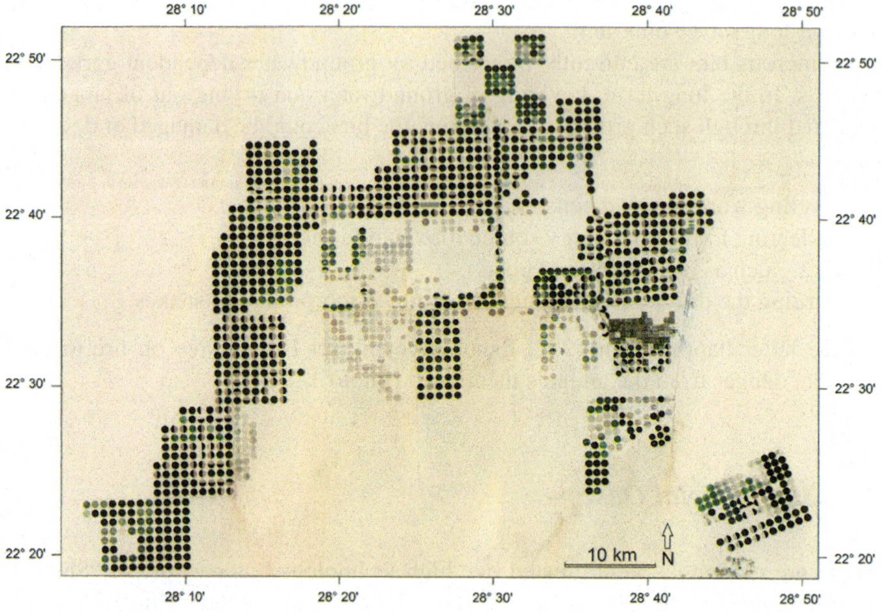


**Fig. 13.2** Satellite image from LANDSAT 8 from February 26, 2017, shows numerous circular fields on the Selima Serir Plain (see Fig. 3)

**Photo 13.1** Wet rice field with white herons. Wet rice is often grown as a soil improver to leach salt out of the soil

**Photo 13.2** Dom palms (*Hyphaene thebaica*) are widespread in the Kharga–Baris oases chain. The picture shows dom palms near a dry artesian water well

**Photo 13.3** Evaporation pond surrounded by tamarisks and heavily saline soil near Bulaq

**Photo 13.4** Termite nest near Bulaq

**Photo 13.5** Larva of the "antlion"

**Photo 13.6** Stick insect (*Carausius morosus*) sighted near Bulaq

**Photo 13.7** Bir Safsaf with reeds and palm trees in the lowest part of the depression. Date palms on the left and dom palms on the right

**Photo 13.8** Stipagrostis and tamarisk bushes are found on higher areas around the birs

**Photo 13.9** As the distance to the groundwater increases, *Alhagi mannifera* = manna tree (in the foreground) and tamarisk are found

**Photo 13.10** Acacia mounds (*Acacia ehrenbergiana*) are found at sites with deep groundwater level

**Photo 13.11** Similar to the Acacia, *Ziziphus spina-christi* occurs at sites with deep groundwater level

**Photo 13.12** Accumulation of plant residues creates in the course of time biogenic hills

**Photo 13.13** Biogenic mounds out of tamarisk remains with a height of approx. 15 m are found near Bir Kiseiba

**Photo 13.14** At night at Bir Tarfawi, we occasionally had a visit from the fennec, who was looking for our food remains (*photo* H. Kehl)

**Photo 13.15** Bir Tarfawi was set on fire by a group of visitors (in 1982) in order to avoid possible danger from scorpions and snakes

## 13.2 Episodic Autochthonous Ecosystem

This ecosystem is independent from groundwater and depends only on the rare episodic rains and, above all, from relief. The small amount of episodic precipitation only enables plant growth where rainwater accumulates from a large surface through lateral influx into a small volume of soil-root-zone (Photos 13.16, 13.17 and 13.18) and at the same time is protected from evaporation by an aeolian sand cover and/or in shady places (Alaily et al. 1987b). Best soil for such an ecosystem is a sandy-loam to loam with a sand cover from 2 to 3 cm thick. Suitable sites for this ecosystem are shallow depressions (especially on playas); water channels (especially those that are interrupted by barrier dunes) in wadis and in the foreland of escarpments (Gilf Kebir, north and east plateaus), also at dune base, as well as rock cracks and pockets filled with soil material (Photo 13.19).

**Photo 13.16** Dead plants in a shallow depression in which rainwater could be collected from a large area through lateral inflow

**Photo 13.17** Barrier dune across a dry water channel. Plants can be seen on the side of the dune where water was collected

**Photo 13.18** Plants are sometimes found in dry water channels of wadis

**Photo 13.19** Plants have been observed on the northern slope of a basalt hill. The roots are found in cracks filled with soil material

Above all, the amount of water stored in the soil and the plant root system determine the plant lifespan. The ecosystem comes to an end when the stored water in the soil is used up by the plants. This can be a few weeks to few years. During our expeditions from 1981 to 1986, almost all plants in the episodic ecosystem were dead and the available water of the soil was completely consumed.

During the last 200 years, the Gilf Kebir may have had an episodic wet period. This is indicated by a buried and charred tamarisk tree found behind a barrier dune in the upper reaches of Wadi Maftuh. This wet period probably lasted 20–30 years. After the death of the tree, a special ecotope was created. We noticed that in the upper part of the trunk there were many living fine root hairs. The root hairs were from dead grasses that could be seen on the soil surface. In contrast to the soil mineral, the tree trunk contained some water that was noticed as condensation on the walls of the sample bag. In the laboratory, plant's available water was measured in the charred wood of the tamarisk tree and in the surrounding soil mineral. A 8–10% water was measured in the charred wood versus < 2% in the soil mineral. From this observation, it can be concluded that due to the high available water capacity of the charred wood compared to the surrounding sand, after the tree died, enough water was stored in the charred wood to allow the growth of some grasses for a short time.

Only four green plants have been observed in the wadis of Gilf Kebir. In 1982, an Acacia tree (*Acacia raddiana*) with green buds was discovered in Wadi Contrast (Photo 13.20); in the same year, *Fagonia arabica* with flowers was found in Wadi Mashi (Photo 13.21); in 1986, *Salsola baryosma* and *Zilla spinosa* were found in Wadi Maftuh, the latter with blossoms (Photos 13.22 and 13.23).

**Photo 13.20** *Acacia raddiana* with green buds found in Wadi Contrast in the north of Gilf Kebir

**Photo 13.21** *Fagonia arabica* with flowers in the upper reaches of Wadi Mashi in the north of Gilf Kebir

**Photo 13.22** *Salsola baryosma* with green leaves in Wadi Maftuh in the east of Gilf Kebir

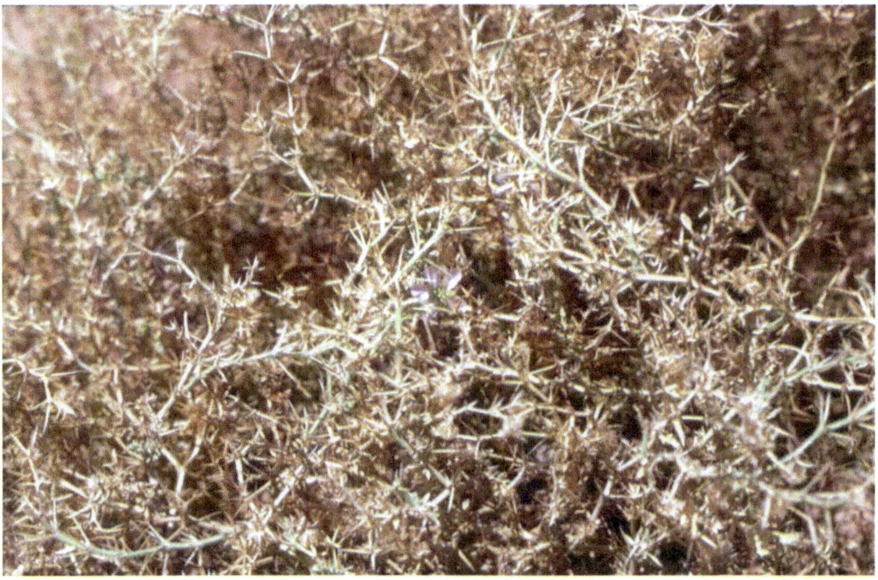

**Photo 13.23** *Zilla spinosa* with flowers (center) in Wadi Maftuh in the east of Gilf Kebir

A total of around 15 plant species in the Gilf Kebir area have been identified by Bornkamm and Kehl (Alaily et al. 1987a, Kehl and Bornkamm 1993). These are: *Acacia raddiana*, *Salsola baryosma*, *Stipagrostis acutiflora*, *Fagonia arabica*, *Zilla spinosa*, *Citrullus colocynthis*, *Astragalus vogelii*, *Trichodesma africanum*, *Rumex vesicarius*, *Schouwia thebaica*, *Onobrychis cristaonia-galli*, *Panicum turgidum*, *Monsonia nivea*, *Prasium majus,* and *Setaria ambigua*. In 1986, a fully buried tamarisk tree was found in Wadi Maftuh.

In the north and east of the investigated region, the following species have been identified by our botanists on the limestone plateau, its escarpment, and especially in its wadis: *Salsola baryosma, Fagonia indica, Astragalus vogelii,* and *Morettia Phileana* (Alaily et al. 1987a). In this area, most plants were dead. It can be assumed that after a rain event, a larger number of plant species will appear in this ecosystem. During our expedition in November 1982, green tamarisk hills and *Alhagi manifera* stocks were observed on a playa in an end pan in the foreland of the Abu Tartur Plateau (east of Dakhla, Photos 13.24 and 13.25). If there is no rain in the following years, the lifespan of the plants and animals of this location will depend mainly on the amount of available water stored in the soil.

**Photo 13.24** In November 1982, an area with green plants in an end pan on playa sediment was observed in the foreland of Abu Tartur escarpment

**Photo 13.25** *Alhagi manifera* with green components in an episodic autochthonous ecosystem in the foreland of Abu Tartur escarpment

In the episodic ecosystem when the plants die, the animals either have to emigrate, die, or live from imported food. In episodic ecosystems that are close to oases, as is the case for example in the north and east of the investigated region, most animals can immigrate to the oases when the plants die. It is similar in northern Egypt in the transition area between semi-arid and fully arid. There is also an episodic ecosystem. After the plants die, large animals emigrate toward the coast. Small animals such as, e.g., snails cannot overcome the distance to the semi-arid area (where food is still available) and therefore have to die (Photos 13.26). In the area of Gilf Kebir, all oases or other permanent ecosystems are relatively far away (several kilometers), and therefore, it is not possible for animals to immigrate to new feeding areas. In Gilf Kebir, when the plants die, so do the animals. The dead animals dry up and are mummified. On our expeditions, several mummified mane sheep (*Ammotragus lervia*) (Photo 13.27) and remnants of a termite nest (Photo 13.28) were found in the area of Gilf Kebir. Four mummified mane sheep were found in Wadi Maftuh alone. In addition, in almost all wadis, we can see animal trails on the slopes.

**Photo 13.26** In northern Egypt, the transition area between semi-arid and full arid zones represents an episodic ecosystem. The picture shows a dead plant surrounded by dead snails that could not immigrate during a dry period to a better ecosystem

**Photo 13.27** During the last wet period, mane sheep lived in Gilf Kebir. Under the present climate, the mane sheeps are mummified (*photo* H. KEHL)

**Photo 13.28** Remnants of a termite nest in Wadi Bakht in Gilf Kebir

The mane sheep could have come from the Tibesti or the Owaynat Mountains. This happened at a time with higher precipitation, within the past 200 years (see Chap. 2, last episodic wet phase). At that time, the Gilf Kebir Plateau was connected to the neighboring mountains by a continuous vegetation cover, so that some herbivores could migrate toward Gilf Kebir. With increasing aridity, the plants died first on the plains in the lowlands around the plateau. As a result, the animals were caught in a kind of oasis (in the wadis and on the plateau), which slowly dried out epicentrically with all plants and animals.

With the exception of a white wagtail that was "traveling" and had rested in Wadi Mashi (Gilf Kebir), no live animals were seen in this ecosystem. Only fresh traces of jerboa, fennecs, and snakes (possibly horned viper) have been found almost everywhere. These animals feed on imported organic matter. This means that in dry periods the episodic ecosystem is transformed into an allochthonous ecosystem.

## 13.3 Allochthonous Ecosystem

This ecosystem is the dominant ecosystem of the investigated region in southwest Egypt. The allochthonous ecosystem is characterized by a lack of producers = vegetation (Bornkamm 1987). It covers the area between the Dakhla–Kharga–Baris oases in the northeast, the Bir Tarfawi and Safsaf in the southeast, and the Gilf Kebir in the southwest. Only three dead plants were found in this huge area (Fig. 13.1). It includes the Selima Serir, Atmur El Kabeisch, Bir Sahara, Bir Misaha, and Bir Abu El Hussein as well as all deflation plains, all ergs, and most of the hilly landscapes. The soil in this ecosystem is free of plant seeds (Alaily et al. 1987a). In an experimental farm on the Selima Serir (south of Bir Tarfawi), no wild plants were observed after several years of regular irrigation. The fields remained free from weeds (Photo 13.29). This means that even after a rain event extremely few plants would appear.

Despite the lack of vegetation, animal traces were seen in several locations, especially the fennec, the jerboa (Photos 13.30 and 13.31), and snakes (Photo 13.32). In addition, two nests of falcons with chicks (Photo 13.33a, b) and the flightless Mantis (*Eremiaphila spec.*) were observed in several places (Photo 13.34). Only two lizards were seen in the area (Photo 13.35). The food for all of the animals mentioned is imported into the ecosystem for the last millennia. This is done by wind, which transports straw, seeds, and insects and deposits them on the leeward of dunes, hills, and under the stones of the soil covers (Photos 13.36 and 13.37). Migratory birds that die on the way also serve as food for secondary consumers in this ecosystem (Photo 13.38). Moreover, food is brought into the ecosystem by people traveling (researchers, tourists, and military), previously by caravans. The allochthonous ecosystem is depending to a certain extent on neighboring ecosystems.

**Photo 13.29** After several years of drip irrigation in an experimental farm, hardly any weeds appeared (photo above), as a comparison, a drip irrigated field in southern Spain near Almeria (photo below)

**Photo 13.30** Traces of jerboa that live under a stone on a Serir plain

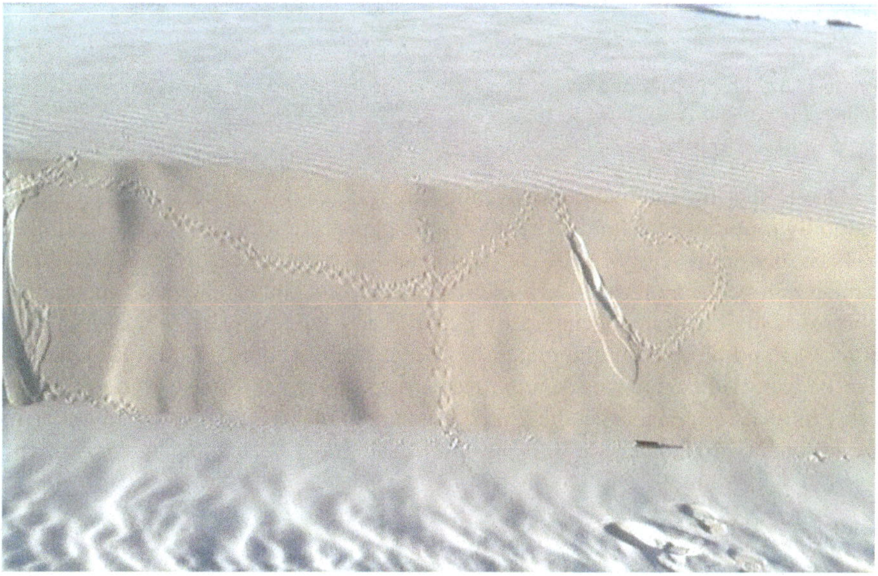

**Photo 13.31** Lee side of a dune with traces of panicked jerboa. At this site, jerboa were feeding on food imported by wind and they were attacked by a snake

**Photo 13.32** Fresh traces of a snake. The traces led to the lee side of a dune, where jerboa had gathered (see photo 194)

Due to the low relative humidity, the carcasses of animals dry quickly and are naturally mummified. Photo 13.39 shows the carcass of a camel that was estimated to have died in the early 1970s and was naturally mummified. Dozens of such dry camels can be seen on the "Darb El Arbain" caravan route. The decomposition of corpses is mainly done mechanically by wind and sand blowers, which grind such carcasses and blow away the grated organic material (Photo 13.40). This material can be exported or deposited on the leeward of objects and thus serve as food for other animals. Occasionally the grated organic animal material is found under the stones on the ground surface. It is powdery, dark gray to black and "greasy".

**Photo 13.33** Lanner falcon (*Falco biarmicus*) and nest with four chicks. The large pile of excrement shows that the nest has been used for several generations

**Photo 13.34** Flightless praying mantis (*Eremiaphila* sp.) can be found everywhere in the investigated region

**Photo 13.35** Green lizard seen near Six Hills

**Photo 13.36** South of the Dakhla and Kharga Oases, on the lee side of the dunes, one often sees organic material that was transported by wind and which serves as food for consumers in the allochthonous ecosystem

**Photo 13.37** Straw and organic material imported by wind, which serves as food for the consumers in the allochthonous ecosystem

**Photo 13.38** Dry bird carcasses serve as food for consumers of the allochthonous ecosystem. This carcass was found next to the falcon nest (on Photo 13.13)

**Photo 13.39** Carcass of a camel that was estimated to have died in the early 1970s and was naturally mummified

**Photo 13.40** Decomposition of animal carcasses is done mechanically by means of wind-sand-blower

Photo 13.41 shows some migratory birds that have been observed in the allochthonic ecosystem. Some rested and then flew their way through the Sahara (Photo 13.41b, c), others were exhausted and seeked help (Photo 13.41d), and some were dead and mummified (Photo 13.41a, e–h). In addition to the birds shown, a lesser gray shrike (*Lanius minor*) was seen near Bir Tarfawi, a chiffchaff (*Phylloscopus collybita*) north of Six Hills, and a dry barn swallow (*Hirundo rustica*) near Bir Misaha.

**Photo 13.41** Some migratory birds observed by the author

# References

Alaily F, Bornkamm R, Kehl H, Renger M (1987a) Evaluation of land in SW-Egypt. Berliner geow Abh (A) 75(2):571–544 (Berlin)
Alaily F, Bornkamm R, Blume H-P, Kehl H, Zielinski H (1987b) Ecological investigations in the Gilf Kebir. Phytocoenologia 15(1):1–20 (Stuttgart-Braunschweig)
Back H-E (1978) Die Pflanzen- und Tierwelt der Sahara. In: Museen Der Stadt Köln: Sahara 10000 Jahre zwischen Weide und Wüste pp 150–157
Bick H (1989) Ökologie. Gustav Fischer Verlag, Stuttgart, p 327
Bornkamm R (1987) Allochthonous ecosystems. Landsc Ecol 1:119–122
Darius F (1989) Wasser-und Nährstoffhaushalt einer unbewohnten Kleinoase Süd-Ägyptens. Diss TU. Berlin, p 83
Ebraheem AM, Garamoon HK, Riad S, Wycisk P, Seif El Nasr AM (2003) Numerical modeling of groundwater resource management options in the East Oweinat area, SW Egypt. Environ Earth Sci 44:433–447
El Sayed E, Dahab K, Ibraheem AM (2004) Hydrogeological and hydrogeochemical aspects of the Nubian Sandstone aquifer in East Oweinat area SW-Egypt. N Jb Geol Paläont Abh 233:121–152 (Stuttgart)
Euroconsult (1983) Regional development plan for New Valley
Kehl H (1987) Zonation and establishment of vegetation in selected uninhabitated Egyptian and Sudanese oases. Catena 14:275–290
Kehl H, Bornkamm R (1993) Landscape ecology and vegetation units of the Western Desert of Egypt. CATENA Supplement 26:155–178 (Cremlingen)
Schneider U (1990) Pflanzenökologische Untersuchungen an Wild- und Kultuepflanzen in der Extremwüste Südägyptens. Diss TU-Berlin, p 292

# Chapter 14
# Land Suitability for Irrigated Agriculture

**Abstract** Based on the electrical conductivity and the SAR value of the groundwater, the water quality can be described as good to medium-good. The evaluation of soil is based on the efforts that need to be made to improve the location and was derived from the following soil properties: salinity, stone density of the soil cover, stone content of the pedon, depth to solid rock, depth of groundwater level, and clay content. Land suitability of the dominant soils as well as the best soil of the various soil associations in southwest Egypt is presented on a land suitability map. It shows that the best soil association is found within the Serir plains. The worst soil association is found on the limestone plateau. To quantify the suitability of soil for irrigated agriculture, the following parameters were further evaluated in detail:

- nutrient content,
- water permeability which is derived from the soil layer with the lowest permeability within 1 m of soil depth,
- available water capacity,
- unavailable water capacity within the topsoil (top 20 cm),
- aeration capacity of the topsoil (top 50 cm).

The quantitative evaluation of soil parameters that are difficult or expensive to improve were given more importance than those that can be improved easily: soil depth and stone content, water permeability, and the salinity of soil were evaluated with a grade (= index) between 0 and 4, while the available water capacity, the nutrient content, the aeration capacity, and the stone content on the soil surface were rated with an index between 0 and 2 or 1.5. The quality index of the best soil is 0 and that of the worst soil is 27. All soils must be fertilized with N and P. Drip irrigation must be combined with basin irrigation. Agroforestry is the best cultivation system for this region.

**Keywords** Suitability orders · Suitability classes · Suitability subclasses · Quality index · Agroforestry · Drip irrigation · Basin irrigation



F. Alaily, *Ecosystems of Extremes*, Advances in Geographical and Environmental Sciences, https://doi.org/10.1007/978-3-031-76910-8_14

The number of inhabitants in Egypt was estimated at 103 million in 2021 and 95 million in 2017 (Wikipedia 2007). This number is currently growing at around 2% annually. In 1970, the population was 33 million, in 1980 about 60 million, and in 2005 about 74 million. The majority of the population lives along the Nile Valley on an area of 40,000 km$^2$. The Nile Valley in Egypt is one of the densest populated areas on earth. In 1970, the population density in the Nile Valley in Egypt was 825 people per km$^2$. Today, it is estimated more than 2400 people per km$^2$. Compared to the Federal Republic of Germany, with a population density of 233 people per km$^2$, the population density of the Nile Valley in Egypt is more than ten times higher.

The Nile Valley, where the bulk of the population lives, forms only 4% of the land surface of Egypt, the remaining 96% of the land surface consists of desert. As a result, agricultural production is limited and the population is depending on food imports. For decades, economists and politicians have been struggling to secure the population with food and to free it from imports. This is mainly done through land reclamation in desert areas. With the help of such projects, food should be secured for the growing population as well as unemployment and poverty should be combated. The Egyptian government is currently using large areas in southwest Egypt for wheat production. For this purpose, groundwater and Nile water from the Nasser Sea have been used. At present, several wells are being drilled on the Serir area in the southern part of the area under investigation (Photos 14.1 and 14.2). In the southeast, the Abu Zed canal is being built, which brings the Nile water from Lake Nasser to the Baris oasis (Photo 14.3). The canal will have a length of 310 km and will allow the irrigation of around 2340 km$^2$ of arable land. The reclaimed farmland is to be settled by 3 million people.

In this chapter, land suitability for agriculture uses in southwest Egypt is evaluated and some ideas on cultivation systems and irrigation methods are presented. Since southwest Egypt is practically free of rain, land evaluation was carried out for irrigated agriculture and for the common crops grown in Egypt. The land suitability for most of the evaluated areas is primarily determined by the quality of the groundwater and soil. The determination of suitability orders, classes, subclasses, and units was mostly done after evaluating the ecological soil parameters.

## 14.1 Groundwater Quality

The hydrochemical properties of groundwater are discussed in detail in Schneider (1986). The locations of the investigated groundwater samples are shown in Fig. 14.1. The quality of groundwater based on some chemical parameters can be derived from Table 14.1 and Fig. 14.2. It should be noted that the groundwater samples from Bir Kiseiba, Bir Abu El Hussein, and Bir Tarfawi come from open water holes. The remaining water samples come from boreholes. The electrical conductivity and sulfate concentrations of the samples from Bir Kiseiba, Bir Abu El Hussein, and Bir Tarfawi are very high compared to the other samples. This is due to the high evaporation from the open water holes. With the exception of sample 1 (with 3280

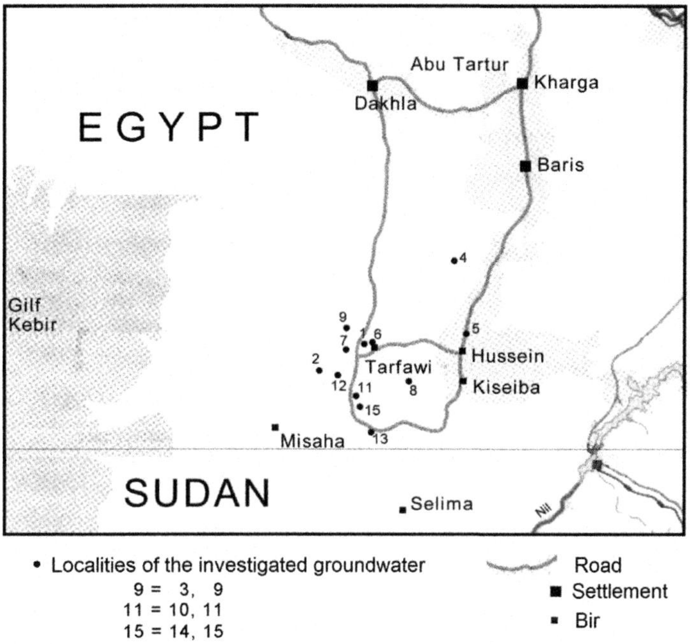


**Fig. 14.1** Locations of the groundwater samples investigated by Schneider (1986)

μS/cm), the electrical conductivity of the drilled wells is between 390 and 1990 μS/cm. The pH value of the birs mentioned is slightly acidic (6.0–6.5) while the water from the boreholes is slightly alkaline to alkaline (7.3–9.4).

In order to assess the quality of water for irrigation and to determine its quality class, in addition to the electrical conductivity values, the sodium adsorption ratio = SAR (United States Salinity Laboratory 1954) was calculated using the following formula:

$$SAR = Na^+/\left(\left(Ca^{2+} + Mg^{2+}\right) : 2\right)^{1/2}$$

(All cations in meq/l)

With the help of the SAR value of water, the danger of alkalinization (solonetization = sodification) of irrigated soils can be estimated. Figure 14.2 shows the SAR values against the electrical conductivities (EC in mS/cm). This indicates that the southernmost samples 10, 11, 13, and 15 have the most favorable values with C2–S1. Otherwise, most samples have values of C3 and between S1 and S2. Samples 1 and 6, which are very close to Bir Tarfawi, show different water qualities, namely C4–S3 for sample 1, C3–S4 for sample 6, and C4–S2 for Bir Tarfawi. The worst water quality with C4–S4 was determined for Bir Abu El Hussein and Bir Kiseiba. In summary, it can be said that the largest number of the investigated groundwater samples has classes C2–C3 and S1–S2. This means that the groundwater has medium to high

**Table 14.1** Cations and anions content (in mg/kg) of groundwater from various localities in southwest Egypt (Schneider 1986)

| No. | $Na^+$ | $K^+$ | $Ca^{++}$ | $Mg^{++}$ | $Cl^-$ | $SO_4^-$ | $HCO_3^-$ |
|---|---|---|---|---|---|---|---|
| 01 | 805.0 | 25.0 | 203.6 | 43.0 | 800.5 | 1004.8 | 392.4 |
| 02 | 184.9 | 8.0 | 22.6 | 14.0 | 162.0 | 24.0 | 348.8 |
| 02a | 182.0 | 8.5 | 12.9 | 12.0 | 170.0 | 1.0 | 286.0 |
| 03 | 138.7 | 5.5 | 22.2 | 10.0 | 206.0 | 14.9 | 148.2 |
| 05 | 286.0 | 10.5 | 9.1 | 6.0 | 369.1 | 90.8 | 87.2 |
| 06 | 399.0 | 15.0 | 11.5 | 7.0 | 481.4 | 99.4 | 222.4 |
| 07 | 222.0 | 11.0 | 89.2 | 24.0 | 329.7 | 87.4 | 322.6 |
| 08 | 280.5 | 10.0 | 37.2 | 7.0 | 422.6 | 64.8 | 109.0 |
| 09 | 164.0 | 5.8 | 15.1 | 10.0 | 233.0 | 19.2 | 106.0 |
| 10 | 36.0 | 4.5 | 27.2 | 10.0 | 42.0 | 48.0 | 101.0 |
| 11 | 56.0 | 4.8 | 39.1 | 15.0 | 71.0 | 96.1 | 103.0 |
| 12 | 135.0 | 8.2 | 19.6 | 20.0 | 175.0 | 105.7 | 73.0 |
| 13 | 120.0 | 6.9 | 27.2 | 15.0 | 136.0 | 96.1 | 132.0 |
| 14 | 154.0 | 4.5 | 70.4 | 19.0 | 179.0 | 240.1 | 112.0 |
| 15 | 28.0 | 2.4 | 40.9 | 10.0 | 48.0 | 48.0 | 106.0 |
| 15a | 160.0 | 4.7 | 55.0 | 21.0 | 270.0 | 172.9 | 62.0 |
| Kiseiba | 1088.0 | 54.8 | 384.0 | 60.0 | 444.0 | 2718.5 | 440.0 |
| Abu El Hussein | 655.0 | 10.1 | 42.8 | 11.0 | 613.0 | 384.2 | 180.0 |
| Tarfawi | 432.0 | 20.7 | 17.0 | 120.0 | 432.0 | 480.3 | 554.0 |

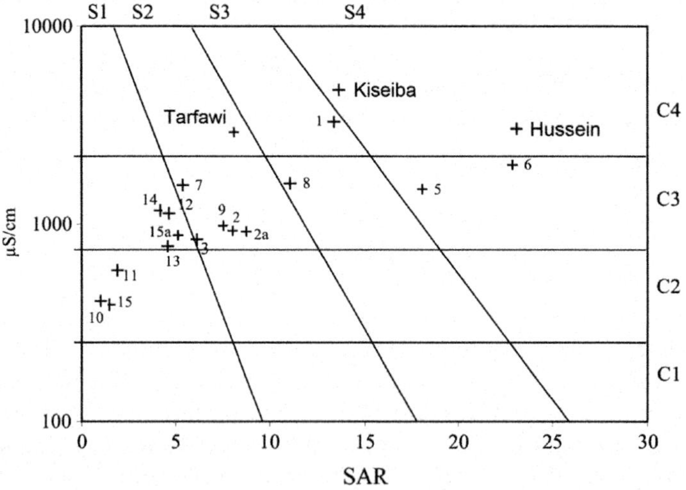


**Fig. 14.2** Position of the samples investigated by Schneider (1986) in the SAR/EC diagram

**Table 14.2** Quality classes and recommended precautions for use of the investigated groundwater from wells and birs

| Grade, sample No | Recommended precautions for use |
|---|---|
| *C2–S1* 10, 15, 11 | Water suitable for plants with moderate salt tolerance and with sufficient washing of the soil |
| *C3–S1* 13, 3, 15a, 12, 14 | Water suitable for plants with good salt tolerance on well-drained soil. Additional salt leaching and occasional application of gypsum in clayey soils are required |
| *C3–S2* 2a, 2, 9, 7 | Water suitable for plants with good salt tolerance on well-drained soil. Additional salt leaching and occasional addition of gypsum to all soils |
| *C3–S3* 8 | Water suitable for plants with good salt tolerance on well-drained soil. Additional salt leaching and regular addition of gypsum are required for all soils |
| *C4–S3* 1 | Water suitable for plants with very good salt tolerance on well-drained soil. Additional increased salt leaching and regular addition of gypsum are required for all soils |
| *C4–S2* Bir Tarfawi | Water suitable for plants with very good salt tolerance on well-drained soil. Additional increased salt leaching and the occasional addition of gypsum are required for all soils |
| *C3–S4* 5, 6 | Unsuitable for irrigation |
| *C4–S4* Bir Kiseiba, Bir Hussein | Unsuitable for irrigation |

salinity and low to medium sodium adsorption ratio. Similar results were obtained by EL Nahry et al. (2010) for a large number of groundwater samples from the same area.

Based on Landon (1991) and Kuntze et al. (P 341, 1994), the precautions for groundwater use depending on its quality class are given in Table 14.2. However, it is to be expected that the water quality of the drilled wells from which no water was extracted at the time of the measurements will improve after a certain extraction time (Schneider 1986).

## 14.2 Land Evaluation for Irrigated Agriculture

In order to make an accurate evaluation of land suitability for irrigated agriculture, the ecosystem dynamics must be known and understood. Land use should be in harmony with the ecosystem and adapted to the environmental conditions. For the investigated region in southwest Egypt, land suitability assessment was determined on the basis of the following factors:

1. The evaporation potential is very high and the loss of surface water is high and rapid, while water stored in the subsoil (> 50 cm depth) evaporates very slowly (Vahrson et al. 1986 and Alaily et al. 1987a).
2. The evaporation of water leads to soil salinization, especially in soils with a high groundwater level such as Gleysols and soils with high unavailable water capacity (Alaily and Blume 1983) and low water permeability such as clay-rich soils.

This means that soils with more than 50 cm depth and high-water permeability are considered well suited for agriculture. Such soils are sandy despite its low nutrient content and low available water capacity. This could be compensated by a large rooting space which is determined by soil depth and its stone content. In addition, both negative aspects of sandy soils can be improved through artificial fertilization and an appropriate irrigation system. This is easier to do than desalinate the soil.

The present land evaluation system is carried out at the exploration level to create maps on a small scale (1: 1,000,000). Therefore, the microclimate and the relief could not be considered. It should be noted that the development of the different soil types is influenced by their relief positions and can thus indirectly reflect the influence of the relief on land use. In addition, the terrain morphology and its influence on land use can be derived from the landscape type.

### *14.2.1 Qualitative Land Evaluation*

Based on the "Framework for Land Evaluation" of FAO-UNESCO (1976), the soils in the investigated region were evaluated for irrigated farming. Soils of areas which are not relevant for use (NR) and therefore meaningless were not evaluated (Table 14.3). Such areas consist of Lithosol or are endangered by wind erosion or migratory dunes. The soils of the remaining areas were then divided into two orders: suitable (S) and unsuitable (N). The orders S and N are differentiated according to their depth and/or stone content (root space). Both properties cannot be improved with the current techniques. The orders S and N were then divided into two to three classes depending on the current degree of suitability of their soils (FAO-UNESCO 1976). This is based on the efforts that need to be made to improve the location.

The classes are derived from the following soil properties:

- salinity,
- stone content density of the soil cover,
- depth of groundwater level, and
- clay content.

The soil properties mentioned above were derived from the soil unit, phases, and texture. In this way, the S and N orders were divided into five classes (Table 14.3). Finally, the classes were divided into subclasses giving more details on the limiting

**Table 14.3** Characteristics of land use suitability classes (FAO-UNESCO 1976)

| Classes | Land use restrictions or degree of suitability |
|---|---|
| S1 | Land without any significant restrictions |
| S2 | Land with low usage restrictions |
| S3 | Land with severe usage restrictions that can be remedied |
| N1 | Land with severe usage restrictions that cannot be remedied and is currently usable |
| N2 | Land with severe usage restrictions that cannot be remedied and currently cannot be used |
| NR | Land that cannot be used with current techniques |

factors (Table 14.4). The subclasses are marked with small letters, from which the causes of the land use limitation can be derived.

High stone content in the soil cover interferes with seed germination, and high salt content has a toxic effect on plants. In such locations, it is necessary to remove the stone cover and wash out the salt. Both actions can be performed using simple methods. Land with such characteristics was therefore classified as S2. Land with soils that have high salinity and with small root space or have petrosaline horizons has been classified as N2.

Soils with high groundwater levels, such as Gleysol, are poor in oxygen in the root space and highly endangered by salinity. Such soils are found in depression. Soils rich in clay are also endangered by salinization because of their low water permeability and high unavailable water capacity. Such soils can be improved by a suitable drainage system. The construction and maintenance of drainage systems

**Table 14.4** Determination key for land use suitability subclasses

| Class | Subclass | Soil property or "phase" |
|---|---|---|
| S1 | | |
| S2 | S2s<br>S2z<br>S2sz | "Stony"<br>"Saline"<br>"Saline" and "stony" |
| S3 | S3u<br>S3g<br>S3m | Clayey (LT-T)<br>Gleysol<br>"Petrosaline" or "petrogypsic" or "petrocalcic"- Horizon on the soil surface |
| N1 | N1x<br>N1l<br>N1m<br>N1xl<br>N1gx oder l | "Petric"<br>"Lithic"<br>"Petrocalcic" or "petrogypsic"<br>"Petric" and "lithic"<br>"Petric" and/or "lithic" Gleysol |
| N2 | N2x<br>N2l<br>N2xl<br>N2gz | "Petric" and "saline"<br>"Lithic" and "saline"<br>"Petric", "lithic", and "saline"<br>"Petric" and/or "Lithic Gleyic Solonchak" |
| NR | | Lithosols and all soils of erg landscapes |

entails high and permanent costs. In addition, the physical properties of the top soil (approx. the top 20 cm) are to be improved by organic fertilization or sand admixture. Therefore, land with a high ground water table and with clay-rich soil was classified as S3.

Areas having soils with small root space (suitability order N) and high groundwater level and/or high clay content were classified with N2.

The S1 class represents areas with soils that do not have any of the above mentioned properties.

Areas with N1 class have soils with small root space or a petrocalcic or petrogypsic horizon in the topsoil. The latter can be broken by deep plowing.

According this classification, areas of the S2 class can be upgraded to the S1 class and areas of the N2 class can be upgraded to N1. Table 14.5 shows the number of quality classes per soil unit. This shows that Lithosol, Vertisol, Gleyic Solonchak, and recent Gleysol are each put in one quality class, namely NR or S3. This is due to the extremely shallow depth of the soils or the high clay content. Both soil properties can be derived from the definition of the soil unit. The other soil units can be assigned to different land suitability classes depending on their ecological parameters. The Haplic Yermosols have the broadest range of suitability classes (see also Fig. 14.4).

The land suitability map of southwest Egypt on a scale of 1: 1,000,000 (Alaily 1990) was derived from the soil association map. Since every soil association unit on the soil association map comprises several soil units and land forms, the land suitability for each map unit must be relatively heterogeneous. In order to facilitate land use planning, only the dominant and the best soils of the soil associations were evaluated and presented on the land suitability map. The determined land suitability

**Table 14.5** Number of land suitability classes per soil unit (n = total number)

| Soil unit | S1 | S2 | S3 | N1 | N2 | NR | n |
|---|---|---|---|---|---|---|---|
| Calcic Yermosol | 6 | 4 | 1 | 3 | 3 | 0 | 17 |
| Gypsic Yermosol | 0 | 1 | 0 | 3 | 2 | 0 | 6 |
| Haplic Yermosol | 7 | 4 | 1 | 10 | 5 | 0 | 27 |
| Cambic Arenosol | 15 | 5 | 0 | 0 | 1 | 1 | 22 |
| Calcaric Regosol | 0 | 0 | 0 | 0 | 0 | 1 | 1 |
| Eutric Regosol | 2 | 0 | 0 | 1 | 3 | 0 | 5 |
| Eutric Fluvisol | 0 | 2 | 0 | 1 | 1 | 0 | 4 |
| Lithosol | 0 | 0 | 0 | 0 | 1 | 20 | 21 |
| Orthic Solonchak | 0 | 21 | 2 | 1 | 34 | 0 | 58 |
| Gleyic Solonchak | 0 | 0 | 7 | 0 | 0 | 0 | 7 |
| Eutric Gleysol | 1[a] | 0 | 1 | 0 | 0 | 0 | 2 |
| Chromic Vertisol | 0 | 0 | 5 | 0 | 0 | 0 | 5 |
| Takyric Yermosol | 2 | 0 | 0 | 0 | 0 | 0 | 2 |
| n | 35 | 40 | 12 | 19 | 50 | 22 | 177 |

[a]Relict Gleysol

classes and subclasses of the dominant soils as well as the best soil of the various soil associations in southwest Egypt is presented in Table 14.6 and on the land suitability map (Fig. 14.3). They show that the best soil association is found within the Serire plains; its dominant soil is rated S1. The worst soil association is found on the limestone plateau; its best soil was rated N2xl.

On the land suitability map of southwest Egypt (Alaily 1990), the map units are numbered from 1 to 17, with unit 1 having the best land suitability with S1 subclass and 17 the worst with NR. Unit 1 occupies about 30% of the investigated region and characterizes most of the Serir plains and wadis of Gilf Kebir. The map unit 17 characterizes the erg landscapes, which occupy about 25% of the investigated region. The Dakhla–Kharga–Baris oases have the map unit 4. Their dominant soils were rated S3tz and their best soils S1. The dominant and the best soil of the limestone plateau north of Dakhla–Kharga and east of Kharga–Baris both have the map unit 12 and were rated with a subclass N2xl. The deflation plains and hilly landscapes are very heterogeneous in terms of their land suitability: their dominant soils have been rated with N or NR and their best soils with S1.

### 14.2.2 Quantitative Soil Evaluation

For the quantitative assessment of soil fertility, the following plant growth factors must be recorded: root penetration, available water capacity, aeration capacity, available nutrient content, the buffering capacity for pollutants, risk of salinization, and workability of the soil. The present quantitative assessment of soil suitability for irrigated agriculture was based on the work of Storie (1964) and Petermann and Blume (1979). Due to the high pH values of the soils, their buffering capacity was only assessed qualitatively and as "good".

To quantify the suitability of soil for irrigated agriculture, the soil parameters used to determine the order, classes, and subclasses as well as the following parameters were further evaluated in detail:

- nutrient content,
- water permeability which is derived from the soil layer with the lowest permeability within 1 m of soil depth,
- available water capacity,
- unavailable water capacity within the topsoil (top 20 cm),
- aeration capacity of the topsoil (top 50 cm).

The quantitative evaluation of soil parameters that are difficult or expensive to improve were given more importance than those that can be improved easily. Accordance to this point of view the root space (soil depth and stone content), water permeability and the salinity of soil were evaluated with a grade (= index) between 0 and 4, while the available water capacity, the nutrient content, the aeration capacity, and the stone content on the soil surface were rated with a grade between 0 and 2 or 1.5 (Table 14.7).

**Table 14.6** Land suitability classes and subclasses of dominant and best soils of the soil associations in southwest Egypt

| Soil assoc | Dominant soil | | Best soil | |
|---|---|---|---|---|
| | Soil unit | Suitability subclass | Soil unit | Suitability subclass |
| *Oasis landscape* | | | | |
| Zo6 c | Zo<br>Vc | S2z<br>S3uz | Qc | S1 |
| *Erg landscape* | | | | |
| D a | D | NR | Qc | S1 |
| *Serir landscape* | | | | |
| Qc2 a | Qc | S1 | Qc | S1 |
| | Re | S1 | | |
| Rc1 a | Rc | S1 | Qc | S1 |
| | Qc | S1 | | |
| Yk1 f | Yk | N2x | Yk | N2x |
| Re1 a | Re | S1 | Re | S1 |
| Yk2 a | Yk | S1 | Re | S1 |
| Zo5 d | Zo | N2x | Re | S1 |
| Zo8 f | Zo | N2x | Re | S1 |
| Zo9 e | Zo | N2zm | Re | S1 |
| Yh4 c | Yh | S2z | Yh | S2z |
| | Zo | S2z | | |
| *Limestone plateau, slopes, and wadis* | | | | |
| Zo1 j | Zo | N2xl | Zo | N2xl |
| Zo3 i | Zo | N2x | Yt | S1 |
| *Sandstone plateau, slopes, and wadis* | | | | |
| I1 j | I | NR | Yh | N1xl |
| Zo2 j | Zo | N2xl | Yh | N1xl |
| Yh1 j | Yh | N2xl | Yh | S1 |
| Qc1 b | Qc | S1 | Qc | S1 |
| *Hilly landscape* | | | | |
| Yh2 g | Yh | N1xl | Yt | S1 |
| I5 j | I | NR | Qc | S1 |
| I4 g | I | NR | Qc | S1 |
| I3 k | I | NR | Vc | S3u |
| I2 g | I | NR | Qc | S1 |
| I7 g | I | NR | Yh | S1 |
| *Deflation plain* | | | | |
| Yh3 g | Yh | N1xl | Yh | Yh |

(continued)

**Table 14.6** (continued)

| Soil assoc | Dominant soil | | Best soil | |
|---|---|---|---|---|
| Yyl k | Yy | N2xl | Zo | Zo |
| Zo4 k | Zo | N2xl | Zo | Zo |
| Zo7 | Zo | S2sz | Yk | Yk |
| | Yk | S2sz | | |
| I6 g | I | NR | Zo | Zo |

Symbols of the limiting factors of the quality subclasses:
m: soil crust > 10 cm thick, l: soil depth, x: stone content, s: stone content of the soil cover, u: unavailable water capacity, z: salinity.

The available water capacity, unavailable water capacity, aeration capacity, and water permeability of the investigated soils were derived from the soil texture and soil density and according to Schlichting and Blume (1966), Schlichting et al. (1995), and Arbeitsgruppe Bodenkunde (1982). The nutrient content was derived from the soil texture and the pH value. With the data of the ecological soil parameters and by means of Table 14.7, the quality indexes of the soils were determined. The sum of the quality indexes of all soil parameters gives the suitability value of the soil for irrigated agriculture. As the value increases, the soil suitability for cultivation decreases. According to Table 14.7, the worst soil should have a quality index of 27 and the best soil 0. Using the example of dominant soils of the soil associations, the estimated ecological parameters and the calculated soil suitability values were determined and presented in Table 14.8. In order to make a precise land suitability decision, the quality index in Table 14.8 is written left from the subclass symbol. In this way, in addition to the suitability value, the degree of usage restriction and its causes can be derived from the abbreviations. The quality index and the subclass together represent the land suitability unit of the present land evaluation inventory. Land suitability unit can be presented mainly on large-scale maps.

The relationship between land suitability subclasses and the determined quality indexes for various soils is shown in Fig. 14.4. It can be seen that the quality index increases as the subclass deteriorates, namely in the order S1, S2, S3, N1, and N2. This is understandable, since the same ecological parameters were assessed for both quality values. The quality indexes of subclasses overlap in a relatively large range. This means that a quality index value can occur in different subclasses without showing the causes of the land suitability limitation.

## 14.3 Some Land Use Recommendations

The agricultural use of the investigated region is hampered by several environmental problems. All terrestrial soils are very poor in nitrogen due to lack of humus and relatively high pH values (7–8) and the availability of their phosphates and most of the

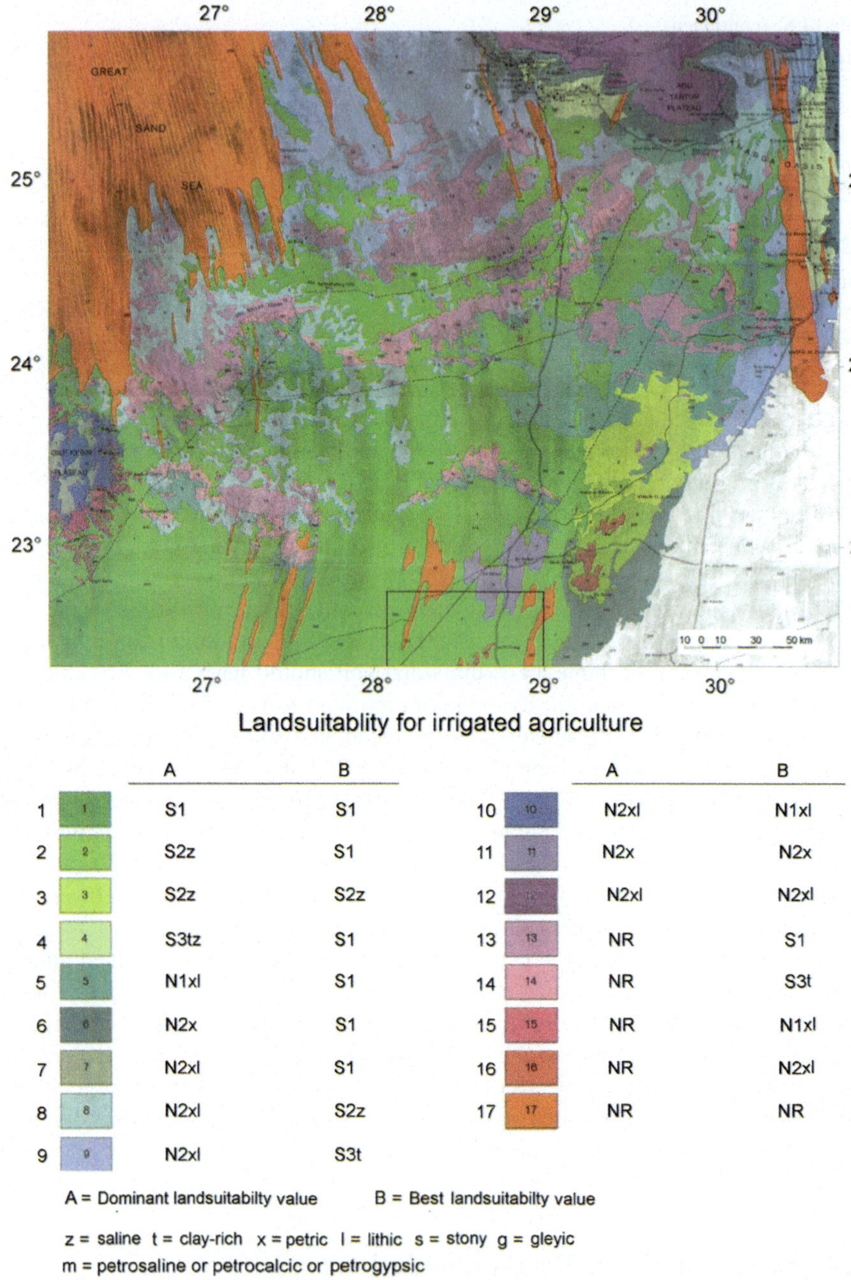


**Fig. 14.3** Land suitability map for irrigated agriculture. Land reclamation projects were implemented within the outlined area in the south (see Fig. 2)

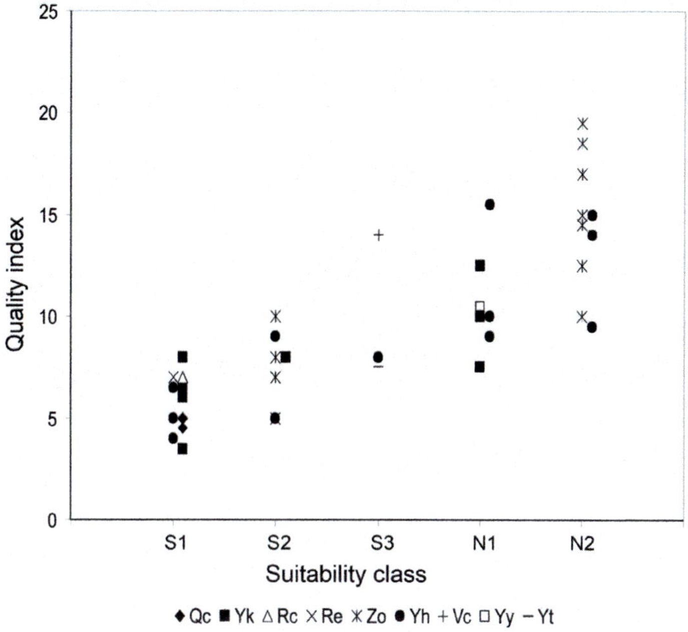


**Fig. 14.4** Relationship between quality indexes, suitability subclasses from some terrestrial soil units

**Photo 14.1** Dozens of water wells are being drilled on the Serir plains in the south of the investigated region

**Photo 14.2** Huge trucks are used to transport the drill derrick

**Photo 14.3** Abu Zed Canal is supposed to deliver Nile water to the oases Baris and Kharga

**Table 14.7** Quality range and quality index of ecological soil parameters

| Soil properties and symbols | Quality index | | | | | | | |
|---|---|---|---|---|---|---|---|---|
| | 0 | 0.5 | 1.0 | 1.5 | 2.0 | 3.0 | 4.0 | |
| Depth (l) | Quality ranges | | | | | | | |
| | > 100 | | 100–50 | | 49–30 | 29–10 | < 10 | cm |
| Stone content (x) | < 5 | | 5–9 | | 10–19 | 20–40 | > 40 | Vol. % |
| uWC (u) | < 6 | | 6–11 | | 12–19 | 20–30 | > 30 | mm in 20 cm Depth |
| Water permeability (p) | > 100 | | 100–41 | | 40–11 | 10–1 | < 1 | cm/d |
| Salinity [a] (z)<br>[b]<br>[c] | < 2<br>< 4<br>< 8 | | 2–4<br>4–8<br>8–15 | | 5–8<br>9–15<br>16–30 | 9–15<br>16–30<br>31–45 | > 15<br>> 30<br>> 45 | mS/cm |
| aWC (a) | > 200 | 199–140 | 139–90 | 89–50 | < 50 | | | mm in 100 cm depth |
| Nutrient reserves (n) | > 200 | 199–100 | 99–50 | 49–10 | < 10 | | | Val. In $m^2$.m |
| AC (o) | > 12 | 11–7 | 6–3 | < 3 | | | | Vol. % |
| Stone content of soil cover (s) | < 10 | 10–19 | 20–40 | > 40 | | | | Vol. % |

uWC = unavailable water capacity, aWC = available water capacity, AC = air capacity.
Lowest water permeability: [a] < 10 cm/d, [b] 10–40 cm/d and [c] > 40 cm/d.

trace elements are very low; therefore, these soils have to be fertilized regularly with nitrogen and possibly with phosphates as well as some trace elements. Compared to nitrogen, phosphorus is not very mobile in the soil; therefore, P fertilizers should be used in granulated form and, if possible, as row fertilization (fertilizer drilled in to the side of the seed row) or tape fertilization (fertilizer tape below the seed row) or fertilization in pockets. For perennial plant species and permanent crops, the N and P fertilization must be adapted to the temporal requirements of the plants and carried out repeatedly (Pagel et al. 1982). For annual plant species, N and P fertilizers should be put into the soil shortly before or during cultivation in order to meet the usually high requirements of plants in the young stage and to reduce the effect of decreasing solubility of phosphates as well as decreasing leaching or gaseous escape of nitrogen.

Due to the high aridity, all soils are potentially endangered by salinization. The plants are endangered by intense direct sunlight, high soil temperatures, and frequent night frost in winter. In addition, the wind causes the following problems:

- The topsoil is eroded by wind storms.

**Table 14.8** Index values (i) of soil parameters from dominant soils and their suitability values for irrigated agriculture for the various soil associations

| b | | 1 | | X | | u | | p | | z[a] | | a | | n | | 0 | | s | | Land suitability |
|---|---|---|---|---|---|---|---|---|---|---|---|---|---|---|---|---|---|---|---|---|
| unit | Type | cm | i | Vol.% | i | 1/m | i | cm/d | i | mS/cm | i | $1/m^2$ | i | Val/ m” | i | Vol% | i | Vol% | i | |
| *Oasis landscape* | | | | | | | | | | | | | | | | | | | | |
| Zo6c | Zo | 100 | 1 | 8 | 1 | 13.0 | 2 | 17 | 2 | 16.0 | 3 | 156 | 0.5 | 160 | 0.5 | 14 | 0 | 3 | 0 | 10.0 S2z |
| | Vc | 100 | 1 | 0 | 1 | 64.0 | 4 | 1 | 3 | 33.7 | 4 | 130 | 1.0 | 430 | 0 | 5 | 1 | 0 | 0 | 15.0 S3uz |
| *Serir landscape* | | | | | | | | | | | | | | | | | | | | |
| Qc2a | Qc | 100 | 1 | 2 | 0 | 5.9 | 0 | 80 | 1 | 7.8 | 1 | 78 | 1.0 | 34 | 1.5 | 28 | 0 | 1 | 0 | 6.5 SI |
| | Re | 100 | 1 | 0 | 0 | 6.0 | 1 | 100 | 1 | 5.8 | 1 | 70 | 1.5 | 34 | 1.5 | 30 | 0 | 0 | 0 | 7.0 SI |
| Rcla | Rc | 100 | 1 | 0 | 0 | 6.0 | 1 | 100 | 1 | 5.8 | 1 | 70 | 1.5 | 34 | 1.5 | 30 | 0 | 0 | 0 | 7.0 SI |
| | Qc | 100 | 1 | 0 | 0 | 6.0 | 1 | 80 | 1 | 1.7 | 0 | 73 | 1.5 | 34 | 1.5 | 29 | 0 | 0 | 0 | 6.0 SI |
| Yklf | Yk | 100 | 1 | 23 | 3 | 6.2 | 1 | 60 | 1 | 6.6 | 1 | 83 | 1.5 | 39 | 1.5 | 20 | 0 | 19 | 0.5 | 10.5 N2x |
| Rela | Re | 100 | 1 | 0 | 0 | 6.0 | 1 | 100 | 1 | 5.8 | 1 | 70 | 1.5 | 34 | 1.5 | 30 | 0 | 0 | 0 | 7.0 SI |
| Yk2a | Yk | 100 | 1 | 1 | 0 | 6.0 | 1 | 60 | 1 | 6.0 | 1 | 71 | 1.5 | 33 | 1.5 | 29 | 0 | 1 | 0 | 7.0 SI |
| Zo5d | Zo | 100 | 1 | 20 | 3 | 5.7 | | 25 | 2 | 25.0 | 3 | 70 | 1.5 | 40 | 1.5 | 19 | 0 | 4 | 0 | 12.0 N2x |
| Zo8f | Zo | 60 | 1 | 28 | 3 | 9.8 | 1 | 25 | 2 | 13.2 | 2 | 86 | 1.5 | 48 | 1.5 | 12 | 0 | 12 | 0.5 | 12.5 N2x |
| Zo9e | Zo | 100 | 1 | 0 | 1 | 6.2 | 1 | 60 | 1 | 120.0 | 4 | 83 | 1.5 | 38 | 1.5 | 20 | 0 | 100[c] | 1.5 | 12.5 N2zm |
| Yh4c | Yh | 70 | 1 | 6 | 1 | 6.3 | 1 | 20 | 2 | 5.9 | 1 | 81 | 1.5 | 49 | 1.5 | 23 | 0 | 2 | 0 | 9.0 S2z |
| | Zo | 100 | 1 | 2 | 0 | 7.9 | 1 | 60 | 1 | 10.1 | 2 | 107 | 1.0 | 50 | 1.0 | 26 | 0 | 2 | 0 | 7.0 S2z |
| *Limestone plateau with slopes and wadis* | | | | | | | | | | | | | | | | | | | | |
| Zolj | Zo | 48 | 2 | 47 | 4 | 12.0 | 2 | 10 | 3 | 22.7 | 4 | 38 | 2.0 | 35 | 1.5 | 7 | 0.5 | 26 | 1.0 | 20.0 N2xl |
| Zo3i | Zo | 100 | 0 | 26 | 3 | 4.4 | 0 | 30 | 2 | 31.4 | 4 | 82 | 1.5 | 47 | 1.5 | 17 | 0 | 36 | 1.0 | 13.0 N2x |

(continued)

**Table 14.8** (continued)

| b | | 1 | | **X** | | u | | p | | $z^a$ | | a | | n | | 0 | | s | | Land suitability |
|---|---|---|---|---|---|---|---|---|---|---|---|---|---|---|---|---|---|---|---|---|
| unit | Type | cm | i | Vol.% | i | 1/m | i | cm/d | i | mS/cm | i | $1/m^2$ | i | Val/ m" | i | Vol% | i | Vol% | i | |
| *Sandstone plateau with slopes and wadis* | | | | | | | | | | | | | | | | | | | | |
| Hj | I | nd | | | | | | | | | | | | | | | | | | NR |
| Zo2j | Zo | 37 | 2 | 64 | 4 | 5.7 | 0 | 25 | 2 | 14.2 | 2 | 16 | 2.0 | 13 | 1.5 | 5 | 1.0 | 24 | 1.0 | 15.5 N2xl |
| Yhlj | Yh | 40 | 2 | 39 | 3 | 8.7 | 1 | 25 | 2 | 6.2 | 1 | 34 | 2.0 | 25 | 1.5 | 10 | 0.5 | 24 | 1.0 | 14.0 N2xl |
| Qclb | Qc | 100 | 0 | 1 | 0 | 6.0 | 0 | 60 | 1 | 1.2 | 0 | 86 | 1.5 | 40 | 1.5 | 28 | 0 | 1 | 0 | 4.0 SI |
| *Hilly landscapes* | | | | | | | | | | | | | | | | | | | | |
| Yh2g | Yh | 50 | 1 | 32 | 3 | 5.0 | 0 | 100 | 0 | 2.1 | 0 | 23 | 2.0 | 11 | 1.5 | 17 | 0 | 21 | 1.0 | 8.5 Nlxl |
| I5j | I | nd | | | | | | | | | | | | | | | | | | NR |
| b | | 1 | | **X** | | u | | P | | $z^a$ | | a | | n | | o | | s | | Land suitability |
| unit | Type | cm | i | Vol.% | i | lW | i | cm/d | i | mS/cm | i | **\imᶻ** | i | $Val/m^2$ | i | Vol% | i | Vol% | i | |
| I4g | I | nd | | | | | | | | | | | | | | | | | | NR |
| I3k | I | nd | | | | | | | | | | | | | | | | | | NR |
| I2g | I | nd | | | | | | | | | | | | | | | | | | NR |
| I7g | I | nd | | | | | | | | | | | | | | | | | | NR |
| *Deflation plains* | | | | | | | | | | | | | | | | | | | | |
| Yh3g | Yh | 18 | 3 | 26 | 3 | 8.0 | 1 | 25 | 2 | 2.1 | 0 | 20 | 2.0 | 14 | 1.5 | 13 | 0 | 22 | 1.0 | 13.5 Nlxl |
| Yylk | Yy | 20 | 3 | 46 | 4 | 9.5 | 1 | 18 | 2 | 5.0 | 1 | 17 | 2.0 | 17 | 1.5 | 6 | 1.0 | 40 | 1.5 | 17.0 N2xl |
| Zo4k | Zo | 41 | 2 | 20 | 3 | 11.7 | 1 | 25 | 2 | 112.9 | 4 | 48 | 2.0 | 33 | 1.5 | 15 | 0 | 46 | 1.5 | 17.0 N2xl |

(continued)

**Table 14.8** (continued)

| b[b] | | 1 | | X | | u | | P | | z[a] | | a | | n | | o | | s | | Land suitability |
|---|---|---|---|---|---|---|---|---|---|---|---|---|---|---|---|---|---|---|---|---|
| Zo7h | Zo | 100 | 0 | 3 | 0 | 13.1 | 2 | 25 | 2 | 13.1 | 2 | 170 | 0.5 | 128 | 0.5 | 17 | 0 | 28 | 1.0 | 8.0 S2sz |
| | Yk | 100 | 0 | 3 | 0 | 7.5 | 1 | 45 | 1 | 6.5 | 0 | 103 | 1.0 | 50 | 1.0 | 24 | 0 | 24 | 1.0 | 5.0 S2sz |
| I6g | I | nd | | | | | | | | | | | | | | | | | | NR |

[a]Calculated from the salinity and water content in the saturation extract within 100 cm of soil depth
[b]Unit: Map units of the soil association map SW-Egypt, type: Dominant soil unit
[c]Salt crust

- With the eroded soil, seeds, seedlings, and fertilizer are blown away.
- The sand blown by the wind causes mechanical damage to plant leaves and seedlings.
- On the lee side of objects, plants will be buried by deposited sand.
- Water loss by evapotranspiration increases with increasing wind speed.

Because of the special climatic conditions, which are characterized by high aridity, continentality, and wind storms, the best cultivation method appears to be that which has been practiced by the oasis inhabitants for millennia. This system can be described with the modern term "agroforestry". It is based on basin and furrow irrigation and multi-umelu cultivation of crops. The top story is formed by a date palm forest, below that is the crown of small fruit trees (e.g., citrus, guyava, olives, grapes, etc.), and below that the smaller crops (peanuts, tomatoes, onions, corn, wheat, barley, rice, etc.) are planted (Photos 14.4 and 14.5). This cultivation method is considered to be the best for the ecosystem, and it is best practiced by the farmers in the oases. In this system, a pleasant microclimate with higher relative humidity is created. The unproductive evaporation of irrigation water from the topsoil is reduced and thus the risk of soil salinization. In addition, the lowest vegetation story is protected from:

- direct sunlight,
- wind-sand-blower,
- being buried by sand,
- soil erosion, as well as

**Photo 14.4** "Agroforestry" location from a distance, the top story consists of date palms, and the middle story consists of small fruit trees

**Photo 14.5** Young agroforestry fields consisting of date palms, citrus trees, and vegetables (onions) near Bulaq (lower photo by: R. BORNKAMM)

- winter night frost.

The influence of plant diseases is known to be very small in this system without the use of pesticides. In addition, the harvest from the three vegetation stories is often spread over the whole year. The farmers' income is accordingly better distributed over the year compared with the monoculture cultivation systems. However, the described "agroforestry" system does not allow the use of large agricultural machines and is therefore dependent on manual work.

In general, it is known that with increasing water loss through evaporation and decreasing infiltration of the soil, the risk of soil salinization increases. Through the basin irrigation operated by the farmers in the oases, the salts are continuously washed out of the soil. In comparison, when sprinkling with the same amount of water, the proportion of water lost through evaporation must be significantly higher. With drip irrigation, which is considered the most economical irrigation method, no water leaves the rooted soil space. As a result, the soil around the moisted soil space is enriched with salt within a short time under the present climatic conditions of Central East Sahara. Observations in the experimental farm at Tarfawi and near Kalabsha showed that sandy to loamy soils were visibly enriched with salt in less than 6 months (Photo 14.6). In addition, due to the high air temperature, the thin water pipes are quickly clogged through carbonate precipitation and must be replaced continuously in short periods of time. In particular, water pipes that are laying unshaded on the soil surface are additionally strongly heated by direct sunlight, which accelerates carbonate precipitation.

**Photo 14.6** After approximately 6 months of drip irrigation, the soil is visibly enriched with salts near Kalabsha south of Aswan

In the case of basin and furrow irrigation, some of the water from the open channels currently used can be lost through evaporation. This can be avoided by using closed channels or tubes. The water supply to the fields should be done with several siphon tube in such a way that each field (parcel) gets its water volume in a short time. The portion of seepage water that leaves the root space is relatively high with this irrigation form. This can be minimized through controlled watering. In addition, the salts are washed out of the soil with the seepage water. The seepage water is not regarded as lost as it often seeps into the groundwater (aquifer) or is enriched in a secondary groundwater story and from there in the so-called drainage ponds. Therefore, the oasis ecosystem is recommended. In order to optimize the system, improve soil fertility and counteract soil salinization; drainage networks are to be installed to guide the soil seepage water to the drainage ponds in a controlled manner. The density and depth of the drainage tubes, the quality of irrigation water (EC and NAR values), as well as the physical properties and salinity of the soils are to be considered when constructing the drainage network.

Furthermore, drainage ponds could be used for aquaculture and as fish ponds (as long as their salt concentration allows this type of use). They are also considered to be important wetlands and serve as resting places for migratory birds.

## References

Alaily F, Blume H-P (1983) Bodengesellschaft der Vollwüste, Dakhlabecken – Ägypten. Mitteilgn Dtsch Bodenkundl Gesellsch 38:443–449 (Göttingen)

Alaily F (1990) Ecological maps of the Western Desert of Egypt; Land suitability map 1:1000000. TFH-Berlin

Alaily F, Bornkamm R, Kehl H, Renger M (1987a) Evaluation of land in SW-Egypt. Berliner geow Abh (A) 75:(2) 571–544 (Berlin)

Arbeitsgruppe-Bodenkunde (1982) Bodenkundliche Kartieranleitung. 3. Ed. Schweizerbart, Stuttgart

El Nahry AH, Elewa HH. Qaddah AA (2010) Soil and groundwater capability of East Oweinat, Egypt using GIS spatial modeling techniques

FAO-UNESCO (1976) A framework for land evaluation. Soils Bull 32:72 (Rome)

Kuntze H, Roeschmann G, Schwerdtfeger G (1994) Bodenkunde. 5 ed. Eugen Ulmer, p 424, Stuttgart

Landon JR (1991) Booker tropical soil manual. Longman, p 474

Pagel H, Enzmann J, Mutscher H (1982) Pflanzennährstoffe in tropischen Böden – ihre Bestimmung und Bewertung. 1 ed. VEB Deutscher Landwirtschaftsverlag Berlin, p 272

Petermann Th, Blume H-P (1979) Yermosol-Solonchak-Bodenschaft des Fezzan, II. Soziologie und Bewässerungseignung. Mitteilgn Dtsch Bodenkundl Gesellsch 29:811–820 (Göttingen)

Schlichting E, Blume H-P (1966) Bodenkundliches Praktikum. Paul Parey, Hamburg und Berlin, p 209

Schlichting E, Blume H-P, Stahr K (1995) Bodenkundliches Praktikum. 2 ed. Blackwell Wissenschafts-Verlag, Berlin, p 295

Schneider M (1986) Hydrogeologie des Nubischen Aquifersystems am Südrand des Dakhla-Beckens, Südägypten/Nordsudan. Berliner Geow Abh A 71:66 (Berlin)

Storie RE (1964) Soil and land classification for irrigation development. Int Congr Soil Sci (Bucarest)

United States Salinity Laboratory (1954) Diagnosis and improvement of saline and alkali soils. US Department Agriculture Handbook, vol 60, p 160

Vahrson WG, Blume HP, Alaily F (1986) Simulation of episodic rainfalls in the western desert of Egypt. Berliner Geow Abh A 72:55–67, Berlin

Wikipedia (2007) Wikipedia, the free encyclopedia. http://en.wikipedia.org/wiki/Egypt

# Correction to: Geology and Lithology

**Correction to:**
**Chapter 4 in: F. Alaily, *Ecosystems of Extremes*, Advances in Geographical and Environmental Sciences, https://doi.org/10.1007/978-3-031-76910-8_4**

In the original version of the book, Fig. [4.5] is Inadvertently published the same fig 4.7 as in Chap. 4. The erratum chapter and the book have been updated with the change.

The updated version of this chapter can be found at
https://doi.org/10.1007/978-3-031-76910-8_4



F. Alaily, *Ecosystems of Extremes*, Advances in Geographical and Environmental Sciences, https://doi.org/10.1007/978-3-031-76910-8_15

**Photo 4.5** Strips of aeolian sand on a Serir plain south of Abu Ballas Plateau

# Appendix

## Soil Typology

I. Number of soil units of the most widespread Tertiary and Mesozoic parent rock, determined according to the WRB, as well as their relief position (P = plain or hump, S = slope, D = depression).



F. Alaily, *Ecosystems of Extremes*, Advances in Geographical and Environmental Sciences, https://doi.org/10.1007/978-3-031-76910-8

| | Sandstone | | | Clay stone | | | Limestone | | | Granite | | | Quartzite | | | Basalt | | |
|---|---|---|---|---|---|---|---|---|---|---|---|---|---|---|---|---|---|---|
| Soil unit | P | S | D | P | S | D | P | S | D | P | S | D | P | S | D | P | S | D |
| Arenic Gleysol | | | | | | | | | | | | | | | | | | |
| Arenic Regosol | 1 | | | | | | | | | | | | | | | | | |
| Aridic Arenosol | 2 | | 2 | | | | | | | | | 1 | | | | | | |
| Aridic Calcisol | | | | | | | | | | | | | | | | | | |
| Aridic Cambisol | 1 | | 1 | | | | | | | 1 | | | | | | | | |
| Aridic Gypsisol | | | | 1 | | 1 | | | | | | | | | | | | |
| Aridic Leptosol | 2 | | | | | | | | | | | | | | | | | |
| Aridic Regosol | | | | | | 1 | | | | | | 1 | | | | | | |
| Aridic Solonchak | 4 | 1 | 2 | 1 | | 1 | | | | | | | | | | | | |
| Calcaric Arenosol | | | | | | | | | | | | | | | | | | |
| Calcaric Leptosol | 5 | | | | | | | | | | | | | | | 1 | | |
| Calcic Gleysol | | | | | | | | | | | | | | | | | | |
| Calcic Solonchak | | | | | | | | | | | | | | | | | | |
| Endosalic Calcisol | | | | | | | | | | 1 | | 1 | | | | | | 1 |
| Endosalic Cambisol | | | | | | | | | | | | | | | | | | |
| Endosalic Gypsisol | 1 | 1 | | | 1 | | | | | | | | | | | | | |
| Eutric Arenosol | | | | | | | | | | | | | | | | | | |
| Eutric Cambisol | | | | | | | | | | | | | | | | | | |
| Eutric Leptosol | 4 | 2 | 1 | | 1 | | | | | | | | | | | 1 | | |
| Eutric Regosol | | | | | | | | | | | | | | | | | | |
| Fluvic Cambisol | | | | | | | | | | | | | | | | | | |

(continued)

(continued)

| | Sandstone | | | Clay stone | | | Limestone | | | Granite | | | Quartzite | | | Basalt | | |
|---|---|---|---|---|---|---|---|---|---|---|---|---|---|---|---|---|---|---|
| Soil unit | P | S | D | P | S | D | P | S | D | P | S | D | P | S | D | P | S | D |
| Gleyic Solonchak | 2 | | 1 | 2 | | | | | | | | | | | | | | |
| Gypsic Solonchak | 8 | 1 | 1 | 8 | 4 | 1 | 1 | | | 1 | | 1 | | | | 1 | | |
| Gypsic Vertisol | | | | | | | | | | | | | | | | | | |
| Gypsiric Leptosol | 1 | | | | | | | | | | | | | | | | | |
| Gypsiric Regosol | | | | | | | | | | | | | | | | | | |
| Haplic Gypsisol | | 1 | | | | | | | | | | | | | | | | |
| Haplic Solonchak | | | | | | | | | | | | | | | | | | |
| Hypersalic Solonchak | | | | 3 | 1 | 1 | | | | 2 | | | | | | | | |
| Hyperskeletic Leptosol | 1 | | | | | | 1 | 1 | | | | | | | | | | |
| Hypocalcic Calcisol | 1 | | | | | | | | | | | | | | | | | |
| Hyposalic Arenosol | | | | | | | | | | | | | | | | | | |
| Hyposalic Gypsisol | | | | | | | | | | | | | | | | | | |
| Hyposalic Regosol | | | | | | | | | | | | | | | | | | |
| Leptic Calcisol | 1 | | | | | | | | | | | | | | | 1 | | |
| Leptic Cambisol | 1 | | | 1 | | | | | | | | | | | | | | |
| Leptic Gypsisol | 8 | 1 | | 1 | | 1 | | | | | | | | | | | | |
| Leptic Regosol | 9 | | 3 | | | | | | | | | | | | | | | |
| Lithic Leptosol | 17 | 2 | 2 | 11 | 2 | 1 | 1 | 1 | | | | | 1 | | | | | 1 |
| Ochric Solonchak | 2 | | | 1 | | | | | | | | | | | | | | |
| Petric Calcisol | | 1 | | | | | | | | | | | | | | | 1 | |
| Petrosalic Solonchak | | | | | | | | | | | | | | | | | | |

(continued)

(continued)

| | Sandstone | | | Clay stone | | | Limestone | | | Granite | | | Quartzite | | | Basalt | | |
|---|---|---|---|---|---|---|---|---|---|---|---|---|---|---|---|---|---|---|
| Soil unit | P | S | D | P | S | D | P | S | D | P | S | D | P | S | D | P | S | D |
| Rubic Arenosol | | | | | | | | | | | | | | | | | | |
| Salic Histosol | | | | | | | | | | | | | | | | | | |
| Salic Vertisol | | | | | | 1 | | | | | | | | | | | | |
| Takyric Calcisol | | | | | | | | | | | | | | | | | | |
| Yermic Arenosol | | | | | | | | | | | | | | | | | | |
| Yermic Calcisol | | | | | | | | | | | | | | | | | | |
| Yermic Cambisol | | | | | | | | | | | | | | | | | | |
| Yermic Gypsisol | 1 | | | 1 | | | | | | | | | | | | | | |
| Yermic Leptosol | | | | 2 | | | 1 | | | | | | 1 | | | | 1 | |
| Yermic Regosol | | | | | | 1 | | | | | | | | | | | | |
| Yermic Solonchak | 7 | | | 2 | | | | | | 2 | | | | | | | | |

II. Number of soil units of the most widespread quaternary sediments determined according to the WRB and their relief position (P = plain or hump, S = slope, D = depression).

| | Alluvial sand | | | Aeolian sand | | | Playa sediment | | | Fluvial sediment | | | Debris |
|---|---|---|---|---|---|---|---|---|---|---|---|---|---|
| | P | S | D | P | S | D | P | S | D | P | S | D | S |
| Arenic Gleysol | | | | | | 1 | | | | | | 3 | 1 |
| Arenic Regosol | 5 | 1 | 2 | 4 | 2 | | | | 2 | 4 | 2 | 4 | 5 |
| Aridic Arenosol | 7 | | 2 | | 3 | 1 | | | | 3 | 2 | 3 | |
| Aridic Calcisol | 2 | | | | | | | | 1 | 1 | | | |
| Aridic Cambisol | 1 | | | 1 | | | | | 1 | | | 1 | |
| Aridic Gypsisol | | | | | | | | | | 1 | | | |
| Aridic Leptosol | | | | | | | | | | 1 | | | |
| Aridic Regosol | | | | | | | | | | | | | |
| Aridic Solonchak | 1 | 1 | | | | | | | 1 | 1 | | 3 | |
| Calcaric Arenosol | 3 | | | 1 | | | 1 | | | | | | |
| Calcaric Leptosol | | | | | | | | | | | | | |
| Calcic Gleysol | 1 | | | | | | | | | | | | |
| Calcic Solonchak | 3 | | | | | | 1 | | | 2 | | | |
| Endosalic Calcisol | 1 | | 1 | | | | | | | 2 | | 2 | |
| Endosalic Cambisol | | | | | | | 2 | | | 1 | | | |
| Endosalic Gypsisol | 1 | | | | | | | | | 1 | | | 3 |
| Eutric Vertisol | | | | | | | | | 1 | | | | |
| Eutric Arenosol | | | | | | | | | | 1 | | | |
| Eutric Cambisol | | | | | | | 4 | | 1 | | | 1 | |

(continued)

(continued)

| | Alluvial sand | | | Aeolian sand | | | Playa sediment | | | Fluvial sediment | | | Debris |
|---|---|---|---|---|---|---|---|---|---|---|---|---|---|
| | P | S | D | P | S | D | P | S | D | P | S | D | S |
| Eutric Leptosol | | | | | | | | | | 1 | | | |
| Eutric Regosol | | | | | | | 1 | | 1 | | | | |
| Fluvic Cambisol | | | | | | | | | 4 | 1 | | 1 | |
| Gleyic Solonchak | 1 | | 1 | 5 | | 4 | 3 | | 1 | | | | |
| Gypsic Solonchak | 1 | | | 3 | | 1 | 5 | | 2 | 14 | | 1 | 8 |
| Gypsic Vertisol | | | | | | | 1 | | | | | | |
| Gypsiric Leptosol | 1 | | | | | | | | | | | | 1 |
| Gypsiric Regosol | | | 1 | | | | | | | | | | |
| Haplic Gypsisol | | | | | 1 | | | | | | | 2 | |
| Haplic Solonchak | | | | 1 | | 1 | | | | | | | |
| Hypersalic Solonchak | 4 | | | 7 | 1 | 2 | 3 | | 1 | 2 | | 1 | |
| Hyperskeletic Leptosol | | | | | | | | | | | | | |
| Hypocalcic Calcisol | 5 | 2 | | 3 | | | | | | | | | |
| Hyposalic Arenosol | | | | 4 | | | 1 | | | 3 | | | |
| Hyposalic Gypsisol | | | | | | | | | | | | 1 | |
| Hyposalic Regosol | | | | | | | 1 | | | 1 | | | |
| Leptic Calcisol | | | 1 | | | | | | | 1 | | | |
| Leptic Cambisol | 2 | | | | | | | | 1 | 1 | | | 1 |
| Leptic Gypsisol | | | | | | | | | | | 1 | 1 | 1 |
| Leptic Regosol | 4 | | 1 | 1 | 1 | | | | 2 | 6 | 1 | 1 | 1 |

(continued)

(continued)

| | Alluvial sand | | | Aeolian sand | | | Playa sediment | | | Fluvial sediment | | | Debris |
|---|---|---|---|---|---|---|---|---|---|---|---|---|---|
| | P | S | D | P | S | D | P | S | D | P | S | D | S |
| Lithic Leptosol | | | | 3 | | | | | | | | | 2 |
| Ochric Solonchak | 3 | 1 | | 2 | | | | 1 | | | | 1 | |
| Petric Calcisol | 1 | | | 1 | | | | | | | | 1 | |
| Petrosalic Solonchak | 1 | | | | | | | | | 1 | | | |
| Rubic Arenosol | 2 | | 1 | 6 | 1 | 1 | | | | 6 | 1 | 2 | |
| Salic Histosol | | | | | | | | | | | | | |
| Salic Vertisol | | | | | | | 6 | | 1 | 1 | | | |
| Takyric Calcisol | | | | | | | 1 | | | | | | |
| Yermic Arenosol | 2 | | | | 1 | | | | 1 | 2 | 1 | 1 | |
| Yermic Calcisol | | | | | | | 1 | | | | | | |
| Yermic Cambisol | 2 | | | | | | 1 | | 1 | | | | |
| Yermic Gypsisol | | | | | | | | | | | | | |
| Yermic Leptosol | | | | | | | | | 1 | | | | |
| Yermic Regosol | | | | | | | | | | | | | |
| Yermic Solonchak | 1 | | | | | | 1 | | 3 | | | | 4 |

# Locations and Altitude

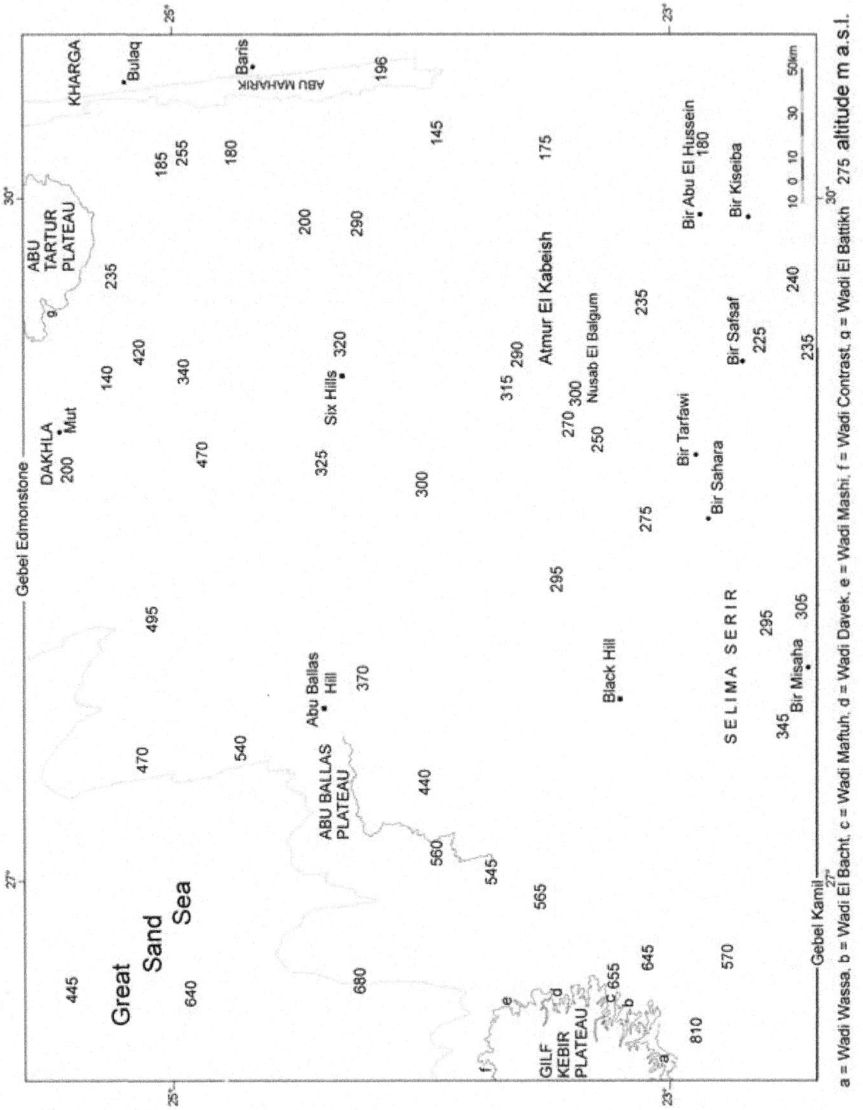


Altitudes and localities mentioned in the text of this Book

# Index




F. Alaily, *Ecosystems of Extremes*, Advances in Geographical and Environmental Sciences, https://doi.org/10.1007/978-3-031-76910-8